NAFANUA

Homalanthus nutans

10 cm.

NAFANUA

Saving
the Samoan
Rain Forest

Paul Alan Cox

Illustrated by Michael Rothman

W. H. Freeman and Company
New York

Cataloging–in–Publication Data

Cox, Paul Alan.
    Nafanua: saving the Samoan rain forest/Paul Alan Cox;
illustrated by Michael Rothman.
        p.   cm.
    ISBN 0-7167-3116-9 (hardcover)
    ISBN 0-7167-3563-6 (pbk.)
    1. Samoans—Ethnobotany. 2. Rain forest conservation—Samoa.
    3. Forest preserves—Samoa. 4. Cox, Paul Alan. I. Title.
    DU819.A2C69   1997
    333.75'16'099613—dc21   97-37379
                                                                            CIP

"The Waking," copyright 1953 by Theodore Roethke. From *The Collected Poems of Theodore Roethke* by Theodore Roethke. Used by permission of Doubleday, a division of Bantam Doubleday Dell Publishing Group, Inc.

Other books by Paul Alan Cox:
Plants, People, and Culture (with Michael Balick)
Islands, Plants, and Polynesians (with Sandra Banack)

Cover and Text Design: Cambraia Magalhaes

Printed in the United States of America

First printing, 1999

Contents

Samoan Pronunciation and Grammar   vii

CHAPTER 1   Voyage   1

CHAPTER 2   Arrival   15

CHAPTER 3   Missionary   35

CHAPTER 4   Return   55

CHAPTER 5   Correspondence   73

CHAPTER 6   Apprehension   95

CHAPTER 7   Crisis   113

CHAPTER 8   Conflict   125

CHAPTER 9   Dreams   137

CHAPTER 10   Responsibility   147

CHAPTER 11   Metamorphosis   163

CHAPTER 12   Hope   175

CHAPTER 13   Deluge   193

CHAPTER 14   Loss   207

CHAPTER 15   Redemption   217

Acknowledgments   227

Sources for Epigraphs   229

Notes on Illustrations   231

Notes   235

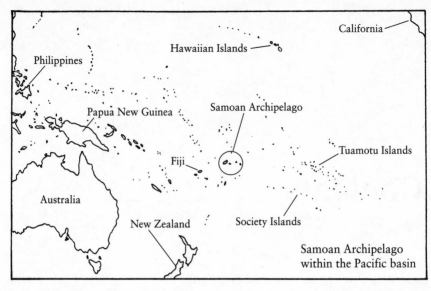

California

Hawaiian Islands

Philippines

Papua New Guinea

Samoan Archipelago

Tuamotu Islands

Fiji

Australia

New Zealand

Society Islands

Samoan Archipelago
within the Pacific basin

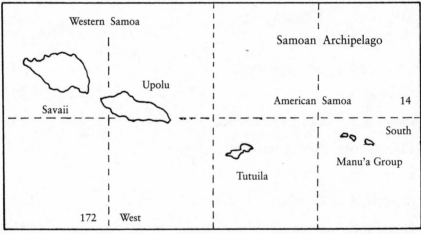

Western Samoa

Samoan Archipelago

Upolu

American Samoa   14

Savaii

South

Manu'a Group

Tutuila

172   West

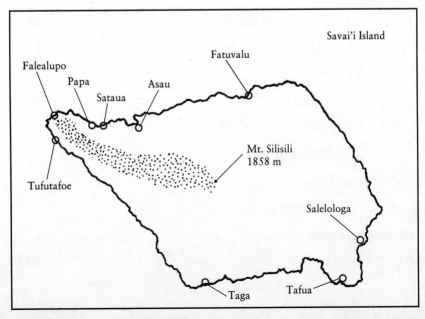

Savai'i Island

Fatuvalu

Falealupo

Papa   Asau

Sataua

Tufutafoe

Mt. Silisili
1858 m

Salelologa

Taga   Tafua

Samoan Pronunciation and Grammar

Samoan vowels are pure, as in French, while the consonants are roughly equivalent to English in pronunciation. The accent is normally placed on the penultimate syllable, but vowels that are to be held out are indicated with a macron as in *susū mai* (welcome). To maintain consistency with previous accounts, however, a macron is not shown over the first vowel of the name of the Samoan goddess of war, Nafanua, which is pronounced "Nah-fah-noo-ah." More nettlesome for foreigners is the "g" sound as in *palagi* (white person), pronounced "pah-LAH-ngee," as in the "ng" sound in "sing along." The glottal stop, indicated by an apostrophe as in *fa'asamoa* (Samoan culture) is similar to the break between words in "Oh-oh." By taking time to pronounce each syllable, Samoan words that at first appear formidable can be easily sounded out.

I have resisted the temptation to anglicize plural Samoan nouns by affixing an "s" (such as in *fales*), since in Samoan the plural can be distinguished from the singular only by context. Thus *fale* can mean either "house" or "houses." However, though Samoan pronouns are genderless, here I translated them into their masculine or feminine English equivalents. Thus the sentence *"Sa aumaia Nafanua ana au tau"* would appear here as "Nafanua brought her war clubs" rather than the more strictly correct "Nafanua brought his/her war clubs."

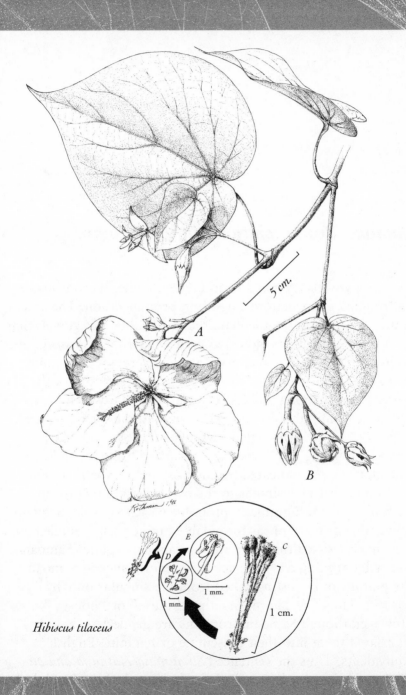

5 cm.

A

B

E

D

1 mm.

C

1 mm.

1 cm.

Hibiscus tilaceus

CHAPTER 1

Voyage

> What gives value to travel is
> fear. It is the fact that, at a
> certain moment, we are so far
> from our own country. We are
> seized by a vague fear, and an
> instinctive desire to go back to
> the protection of old habits. . . .
> Travel, which is like a greater
> and a graver science, brings us
> back to ourselves.
>
> Albert Camus,
> *Notebooks, 1935–1942*

The hospital was spartan. The green walls, metal-frame furniture, and harsh fluorescent light were disorienting, making it impossible to know if it was day or night outside. But none of us noticed the clock or attempted to raise the window shade—our attention was focused on the unconscious woman in the bed.

I had driven my mother to the Salt Lake City hospital two days earlier, well after her metastatic breast cancer had grown resistant to all known forms of chemotherapy. We cared for her at home, but when her agony became unbearable, I arranged for her readmission to the hospital. This would be her last trip, the oncologist cautioned me.

We maintained a constant bedside vigil at the hospital, where she had lapsed into a coma. A small cot was provided for my father, while my brother, sister, and I took shifts. My wife, Barbara, and our children visited briefly. On the third night I saw my mother writhe as if in torment and summoned the nurse for more morphine.

"It isn't time for her dose."

"Dr. Riley said my mother could have whatever she needed, and I don't want her to be in pain. Please give her more." Reluctantly, the nurse complied.

About an hour later, just after my sister Pat arrived, my mother suddenly awakened. She struggled to sit up, but her morphine-induced stupor made it difficult. Finally, half-sitting, she opened her eyes and looked into mine.

"I love you," she said.

"I love you too, mother."

My mother continued to fix her eyes on mine, and, as the breath disappeared from her body, she continued to form the same words with her lips. "I love you. I love you. I love you." And then she was gone.

I held her hand, which soon grew cold. My sister and father wept. After a few minutes, I telephoned my brother, who was heartbroken to learn that he had missed our mother's last moments. I glanced at my watch—5:00 A.M.

I walked through the dark to the parking lot. Silent and alone, I drove the fifty miles back to my home. As the first rays of the sun illuminated the snows on the Oquirrh Mountains to the west, I vowed I would do whatever I could to fight the disease that killed my mother.

The Mulifanua wharf was dark and the air heavy with exhaust fumes. It was an hour before dawn, but already a long line of vehicles snaked down the road, waiting for the ferry to take them from Upolu to Savaii island, a sea journey of thirteen miles. Some were rusted pickup trucks sagging with kava roots, rolls of multipurpose mats, or cooked pigs. Others were Samoan government Land Cruisers bound for forestry offices and village medical clinics. A large white van was emblazoned with the logo of the World Health Organization. But most were logging trucks that had disgorged their rain forest timbers the previous day in the nearby capital city of Apia. The huge and intimidating empty trucks smelled of diesel and creosote, the odor of rain forest death. The logging industry, had, however, given birth to this ferry. Before the Asau sawmill was built on Savaii, transit between the two islands was confined to tiny boats transporting copra—dried coconut meat.

The drivers of the logging trucks were affable—ordinary Samoans lucky enough to win one of the few jobs on Savaii that paid in cash. Yet in the decade since the establishment of the sawmill, these trucks and these drivers and this ferry had already transported from Savaii over 80 percent of its lowland rain forest—perhaps the largest and most species-rich rain forest remaining on any South Pacific island. The rain forest trees had been converted mostly into wooden pallets for the tuna industry and some hardwood timbers for local consumption and export.

The sawmill's destruction of the rain forest had deep implications for the future of Samoa. Many of the forest plants represent botanical outliers, species or genera at the edge of their range, since Samoa is the eastern outpost of plants that trace their evolutionary provenance to Indo-Malaysia. The biological affinities with Indonesia and Malaysia are mirrored in Samoa's thirty-nine resident bird species and three species of native terrestrial mammals—two flying fox species and a species of cave-dwelling bat. As the rain forest dwindled, erosion, flooding, and surface runoff from torrential rains would increase, building up silt in the coral reefs and preventing recharge of subterranean aquifers, which in turn would cause springs to dry up. Indigenous plants of unknown pharmaceutical promise would disappear forever, as would species of animals supported by the island's forests.

Such tropical forests were of tremendous interest to me. I had passed up medical school for graduate training in tropical rain forest biology and ethnobotany, the study of how indigenous peoples use plants. Perhaps I could seek potential anticancer remedies there. Only five months had passed since my mother's death, and only three since I had received a letter from the White House honoring me with a National Science Foundation Presidential Young Investigator Award that would fund any scientific research project I chose to pursue for a five-year period. I arranged an immediate leave of absence from my professorship at Brigham Young University to travel with my family to Samoa for the first year of the award. But I hoped I wasn't too late: relentless logging of the Samoan forest and unavoidable Westernization of the culture meant that the potentially healing plants and the knowledge of how to use them could not survive for long.

I quietly closed the door of our Mitsubishi van so as not to wake the children and walked to the edge of the lagoon. The sea was just beginning to reflect the subtle pink of the clouds. Far beyond the breakers I could see the emerging outline of Savaii, the largest island in Polynesia outside of Hawaii and New Zealand. Underfoot, the shore was littered with broken coral unearthed by recent dredging of the ferry channel. As I switched on my flashlight, I spotted something unusual—a thin, flat brownish object about half the size of a credit card. It was a pottery shard. In my hand it seemed rough and crude, but I could see remnants of ornamentation. Five minutes more of searching along the beach yielded other pieces of what had evidently been a handcrafted pot.

It was during the dredging of the Mulifanua wharf area that the first Lapita pottery in Samoa had been discovered. The dredge line disturbed not only coral and fish, but also the underwater tomb of an ancient village. A team of archaeologists who happened to be working nearby were amazed by the discovery, since there was no history or even oral tradition of pottery manufacture in Samoa. The ornamentation on the shards was similar to prehistoric ceramics first discovered at Lapita village in New Caledonia (eighteen hundred miles from Samoa) and subsequently at other sites in Melanesia, including New Britain and New Ireland (three thousand miles away). The smoke from ancient cooking fires soon burned again in the chemists' laboratories, this time giving radiocarbon dates of 500 B.C. for the Samoan village that had long since subsided beneath the waves. The pottery shards conclusively pointed to Samoa as one of the early centers of Polynesian cultural diffusion. Perhaps the pottery was later rendered unnecessary by the use of fermentation pits for food preservation and pit ovens for food preparation.

I walked back to the car to show the shards to the children. Nine-year-old Emily and seven-year-old Paul Matthew, our two oldest, turned the shards over and over in their hands. Four-year-old Mary and our baby Hillary were just too little and too sleepy to show much interest. Barbara gave the children bananas and passion fruit juice. Together we tried to calm their apprehension about our impending year-long sojourn in Savaii.

When the tropical sun rose, it soon became too hot to remain in the van, even with the doors and windows open. The children elected to don swimming suits and play along the beach next to the wharf. After a few hours, Barbara and I called them to a picnic of baked-potato-like taro and coconuts that we had purchased in the Apia market the day before.

The crowd of Samoans on the wharf, which had been supplemented by numerous deck passengers delivered by bus, was characteristically stoic in waiting for the ferry. As the day became hotter, even the tepid seawater could not cool the children, and all of us became increasingly impatient with the lateness of the ferry. Some Samoans sought respite in the shade beneath the diesel trucks, while others sat quietly on mats spread on the concrete floor of the ferry building. At last, after eight hours in the sun, we drove our vehicle up the steel ramp of the *Limulimutau*. The ferry captain smiled in surprise as I asked for instructions on parking our vehicle. "You speak good Samoan," he said. "Where are you going?"

"Falealupo."

I had chosen Falealupo as a research site by studying maps of Savaii, the largest and least developed island of the Samoan archipelago, searching for villages close to lowland rain forests but as far as possible from Western influences. I was gambling that if there were any herbalists left in the islands, the best might be found in just such a village.

The thirteen-mile trip from Upolu to Savaii on the *Limulimutau* took three hours. Although the sea was choppy, the children quickly fell asleep, as did Barbara. I climbed up on a fifty-five-gallon oil drum next to the rusted blue steel side of the ship. Peering over the rail, I saw the silhouette of Savaii in the distance. I could see the high mountains of the interior, the small volcanic cinder cone of the Tafua peninsula, and the entire land mass brooding as a dark presence on the horizon. As we neared the island, the Samoans on the boat became subdued. I grew apprehensive myself about taking my young family to such a remote and unknown place.

As they colonized other island groups throughout the Pacific, Polynesians retained stories and legends of this place, Savaii. Wherever they encountered such a large island, they christened it with some variant of the sacred name: Havaiki, Hawaii. In most Polynesian legends, the spirits of the dead make their way to where the sun sinks into the sea, just beyond the westernmost point of Savaii, a rocky projection at Falealupo called 'O Le Fāfā. "The disembodied spirit was supposed to retain the exact resemblance of its former self," explained missionary John Stair, who lived in Samoa in the mid-nineteenth century.

> Immediately on leaving its body [the spirit] was believed
> to commence its solitary journey . . . until it reached the
> extreme west point of Savaii, the most westerly island,
> where it finally dived into the ocean and pursued its solitary
> way to the mysterious Fafa.[1]

This belief retains some currency among modern Samoans, and Savaii is still seen to be sacred. In a variant of the mythic search of Orpheus in the underworld for Eurydice, a Samoan urban legend recounts the travails of a woman distraught at her husband's death. She rode the ferry to Savaii, took the bus to Falealupo, and then leapt off the rocks into Pulotu, the underwater world of spirits, to seek her husband. Though Greek and Samoan versions disagree on the sex of the protagonist, there is consensus on the gender of the underworld's ruler: in Samoa Persephone becomes the goddess Nafanua.

But unlike this Samoan Orpheus, I had not come to Samoa to seek an audience with Nafanua. Suddenly there was a commotion at the stern. Several people lunged forward to stop a small boy from being pulled overboard. Behind him in the sea, a marlin flipped and danced on its tail. The boy seemed to be an unwilling participant in a tug of war between the adults on the boat, who grasped his clothing, and the marlin, which had seized the fishing line trailing from a stick in the boy's hand. For a few moments the boy teetered between sea and boat, and then suddenly the line snapped. The boy fell away from Pulotu, back into the arms of the living.

The heat, the rocking of the boat, and the presence of my silent, sleeping family made the voyage seem interminable. Yet the vehicular ferry was far more rapid than the motorized copra boats I used to ride to Savaii two decades earlier. And those copra boats were faster still than the indigenous sailing craft that formerly traversed the channel. Such seacraft carried to Savaii Samoans who sought from Pulotu something more than conversation with deceased spouses—they went to Falealupo to solicit aid in war from the goddess Nafanua. As Captain Wilkes of the U.S. Exploring Expedition wrote of Samoans in 1845, "They acknowledge one great god, . . . but pay less worship to him than to their war god . . . Onafanua [Nafanua]."[2] Nafanua "was the daughter of Saveasiuleo, the god of Pulotu," missionary George Turner recorded in 1884. "She came from Pulotu, the Samoan haedes, when the ruling power was so oppressive."[3] Nafanua freed the Samoan people from foreign oppression, so the legends go, established the first unified Samoan government, and taught the people to protect the forests.

A thatched temple, presided over by chief Auva'a, accepted supplications to Nafanua from both warrior and diplomat. "In war," Turner explained, "all assembled to be sprinkled with Nafanua's cocoa-nut water before going to battle."[4] According to John Stair, "a truce was effected between two armies by the opposing parties mutually agreeing to lay between them Nafanua, one of the national war-gods."[5] Turner explained that "in a case of concealed theft, all the people assembled before the chiefs, and one by one implored vengeance on himself if he was guilty. If all denied, the chiefs wound up the inquiry by shouting out, 'O Nafanua! [Be] compassionate [on] us, let us know who it was, and let death be upon him!'"[6]

Despite Savaii's importance in Samoan religion and politics, early European voyagers largely ignored it. Upolu was frequented by early whalers and adventurers because of its safe harbor near Apia and its large western plain, which provided an abundance of agricultural produce. Savaii, one-and-a-half times the size of Upolu, with interior volcanoes soaring two thousand meters above the sea, was considered too forbidding.

One explorer, however, had a different view. "On Tuesday morning we found ourselves in the straits between two of the largest and most beautiful islands we had yet beheld, having on one side Savaii being two hundred and fifty miles in circumference, and on the other Upolu, which is about two hundred,"[7] wrote John Williams, the first European missionary to the islands, in 1830. Unlike other missionaries operating in the South Pacific, Williams was determined to minimize the impact of European culture on the Polynesians, and so followed the counsel of a Samoan convert "not to commence labours among his countrymen by condemning their canoe-races, their dances, and other amusements."[8] Unfortunately, Williams's sailing vessel, *The Messenger of Peace,* arrived in Samoa in the midst of one of the highest forms of Samoan entertainment: war.

> We were informed that a battle had been fought that very morning, and that the flames we saw were consuming the houses, the plantations, and the bodies of women, children, and infirm people who had fallen into the hands of their sanguinary conquerors.[9]

Williams wisely decided not to initiate his missionary labors until he had obtained permission from the highest chief in Samoa, Malietoa:

> About four o'clock in the afternoon, in a heavy shower of rain, the celebrated old chieftain Malietoa arrived. He appeared about sixty-five years of age, stout, active, and of commanding respect. . . . We expressed our deep regret at finding him engaged in so sanguinary a war, and inquired whether these differences could not be settled amicably, and the dreadful contest terminated. . . . He promised that he would take care there should be no more wars after the present; and that as soon as it was terminated, he would come and place himself under the instruction of the teachers.[10]

Malietoa was true to his word—after he won the war with Upolu, he converted to Christianity, and within two years, most of the inhabitants of the archipelago followed suit, a remarkably rapid and painless process compared with the proselytization of indigenous peoples in other parts of the world. Perhaps this mass conversion was driven as much by respect for Malietoa as by the appeal of the new theology. But one aspect of Samoan legend facilitated the religious transformation of Samoa. Anciently Nafanua told the people that they should join a future kingdom that would come from across the sea. Her unfulfilled prophecy created an empty cultural niche into which John Williams had unwittingly sailed.

Our ferry was now close enough to shore that I could see the large monument at Sapapali'i village commemorating the landing of John Williams. The encounter between Williams and Malietoa forever changed the course of Samoan history. Even today chiefs refer to the period before Williams arrived as "the time when Samoa resided in darkness." Yet as an ethnobotanist, I find the time of darkness, the era of Nafanua, of greater scientific interest than that subsequent to the advent of Christianity. For despite John Williams's precautions, with Western religion came Western culture, and with Western culture came subtle but inexorable erosion of indigenous knowledge systems, a cultural erosion that even I and my family, through our presence, would likely accelerate.

The ferry stopped swaying as we entered calm waters through a channel cut in the reef. Ashore on the Salelologa wharf, a tremendous crowd awaited the ferry's arrival. When we drew close, the steel drive-on ramp that formed the ferry's bow was slowly lowered from its upright position, giving the boat the appearance of a giant landing craft executing a military assault. The ramp was secured to the wharf, and the deck passengers surged toward the waiting buses. As the last deck passengers left, the engines of the empty logging trucks sputtered to life. One by one, belching clouds of diesel, the trucks drove onto Savaii like hungry ravens in search of prey.

We waited our turn and then drove off the ferry, heading toward the narrow road that led inland in the direction of Salelologa. Our van could barely squeeze past the long line of trucks awaiting the return voyage. Unlike the empties that had shared our passage to Savaii, these trucks bound for Apia were weighed down with huge rain forest logs, some exceeding two meters in diameter. Interspersed throughout the line were smaller pickup trucks filled with taro and bananas destined for market. Two policemen in formal blue lavalava (a Polynesian skirt) and white hats watched over the passengers and vehicles jostling for position on the return voyage.

The buses we passed appeared to be strange chimeras of island engineering and diesel truck. Each had sides of wood, with a Plexiglas windshield somehow miraculously inserted into a breadfruit-wood frame. The buses were painted in Seabee blue, or mauve, or whatever tint happened to be in stock that month at the Burns Philp store. But each retained its individuality with florid murals and names garishly painted on the side: 'O le Fetu Ao ("The Morning Star"), 'O le Gata Ula ("The Red Snake"), 'O le Pasi o le Va'a ("The Boat Bus"). Blaring from the inside of the buses were the sounds of Bob Marley, Jim Reeves, or Samoan gospel singing recorded right off the radio on tape that, judging by the clarity of reproduction, must have been manufactured from recycled asphalt. I knew from previous experience that riding such buses is an ordeal not soon forgotten. If you are not deafened by the tape player, asphyxiated by diesel fumes, or crippled by the narrow, hard wooden benches, you still risk, in case of collision, death from splinters.

We drove to the two shops that constitute Salelologa, the only town in Savaii. Entering Burns Philp, I looked across the counter at neatly stacked rows of corned beef and canned mackerel, a few bolts of colorful floral print cloth, and tins of food whose presence in Samoa seemed anomalous: Pears in Syrup (Watties Ltd. of New Zealand); Hunt's Tomato Paste; Goobers Peanut Butter and Jelly Spread, as well as several cellophane packs of Chick-o-Bits, manufactured in Fiji. I gazed reverently into the only refrigerator on the island and asked the shop girl for six cold bottles of Currie's Cordial (Apia). Since the taste of these carbonated beverages is the same regardless of the color or picture on the label, I purchased a random variety of "flavors."

The children, overheated and slightly seasick, were grateful for the cool drinks. We traveled about five miles along the road—the only road—through extensive coconut plantations to the village of Faaala in Palauli district. There, we turned onto the grass in front of a small *fale,* or traditional Samoan house: an oval, open-air structure built on a stone foundation with a peaked thatched roof supported by posts.

"Talofa Tauvao!" I called out to the old man sitting in the *fale.*

"Koki!" he shouted back, calling me by name in Samoan. A big grin filled his face. "You've brought your entire family this time."

I introduced Barbara and the children to the aged chief. Tauvao held the two younger children in turn on his lap, and warmly shook hands with the two older children. He was particularly taken with Paul Matthew's blond hair, which he stroked with his wrinkled brown hand. Had I returned to study the plants of the nearby Tafua peninsula?

"No, Tauvao. We're on our way to Falealupo."

Tauvao's face darkened, and he asked us to stay with him. He feared that our children might get sick from the bad water in Falealupo, and besides, there were plenty of plants in the Faaala forest. Not wishing to pursue his warning further, I changed the subject by asking him to show our children the village turtles.

Tauvao led us across the grass to a freshwater pool next to the sea's edge. Two young Samoan women with long black hair quickly hitched their lavalava above their breasts and gathered the clothes they had been washing. Their embarrassment was quickly overcome by their curiosity about our blond-haired children. Both smiled when I apologized for startling them. The children gazed into the pool at the two green sea turtles lounging near the bottom.

"If you wait quietly for a moment, they'll come up for air," Tauvao explained while I translated.

As if on cue, one of the turtles slowly began to rise. Reaching the surface, it gracefully extended its neck and exhaled. After inhaling, the turtle descended again to the pool bottom.

After bidding Tauvao farewell, we drove west past the Palauli waterfall through coconut groves and banana plantations. Although Barbara had visited Savaii during my doctoral studies, I was looking forward to showing the children the beautiful Taga rain forest that arched like a bower across the road. We rounded a bend and saw long rows of young eucalyptus saplings planted between the burnt husks of rain forest trees on a five-hundred-acre plot. The entire forest had been cut and replaced with eucalyptus, a massive undertaking. I pulled over to ask a passing Samoan what had happened.

"Foreign aid."

We continued to the coast, where the road cut through a long, windswept strip of *Pandanus* forest. These strange-looking trees, held aloft by their stilt roots and adorned with pineapple-shaped fruit, provide the leaves used in Samoan weaving. Over millennia the Polynesians have produced varieties that are planted in neat rows in the villages. None of the improved varieties, though, has the charm of its wild ancestors that stand as sentinels along the coast.

Twenty miles from Tauvao's house, we finally reached Taga village, where a crudely inscribed sign proclaimed the entrance to the Taga blowholes. We turned in. A ten-year-old boy motioned to us to stop. Peering into the car, he counted the number of passengers and in carefully rehearsed English said, "Six *tālā*, please."

I paid the entrance fee, roughly equivalent to three U.S. dollars. When I first came to Samoa, I was irritated by village requests for payment to visit a waterfall or beach. Later, I began to view these requests in a new light. In developed countries, taxes and entrance fees support the preservation and maintenance of national parks, historical monuments, and natural wonders. But in poorer countries such as Western Samoa, it is usually only a single individual, family, or village that takes the responsibility of cleaning and maintaining a beach or waterfall. So the fees, which I once considered an imposition, I now regarded as an opportunity to support grass-roots conservation.

We drove the van down two ruts that extended across a swath of black lava sandwiched between the sea and a field of green

Achrostichum ferns. As we rounded a bend in the road, we heard a loud *whoosh* like a jet engine and saw the eruption of a geyser high into the air.

The children scampered across the lava field to the edge of several large tidal pools. The pools were on the brink of a massive, mile-long lava block whose side dropped precipitously to the sea. There was no reef here, so the waves crashed directly against the lava block, and then were funneled into a series of sea caves. Throughout this network of subterranean catacombs water pressure mounted until geysers of seawater erupted, sometimes ten meters into the sky.

Emily, who had received a camera for her eighth birthday, wanted a photograph of herself and her brother and sisters in front of the blowholes. I positioned the children, and they waited patiently for the eruption. A huge wave headed for the lava block and broke behind them, drenching them with sea spray from the subsequent geyser. I snapped the picture and we left quickly for the car.

After drinking some coconuts we had in the van, we drove back to the main road at Taga village and continued westward toward the Falealupo peninsula. I marveled at the changes that had occurred since I first visited Savaii in 1973. Instead of the rough track that rattled trucks and cars to pieces, smooth pavement now extended over fifty miles around the western end of the island to the sawmill town of Asau. Villages near the sawmill had the glow of electric lights at night, powered by the mill generator. Signs of prosperity increased the closer we drove to Asau: corrugated iron roofing instead of traditional thatch, European-style houses with concrete foundations and louvered glass windows, and new pickups filled with people, mats, and kegs of corned beef. But as I considered what remained of the coastal rain forests, I realized that Savaii had paid a massive price for this prosperity.

In the early 1970s, an American firm named Potlatch built a sawmill and a wharf in Asau with an eye to shipping rain forest timber directly to American and Japanese markets. Unfamiliarity with tropical logging operations, together with ineptness in dealing with Samoan culture, however, led to the sale of the sawmill, first to Australian owners and subsequently to the government of Western Samoa, though the Australians continued to manage daily operations.

I knew of the operation in 1973, but didn't meet anyone involved with it until 1978, when I began research for my doctoral dissertation at Harvard. I needed accommodation in Savaii, so I proposed to one of the Australian managers that I identify plant species for them in return for housing.

A company executive with short pants, white knee socks, and a beer belly looked me up and down.

"You're a Yank, are you?"

"Yes, sir," I replied.

"Are you a greeny?" he asked.

"What does that mean?"

"You know, are you green? Do you want to stop progress, chain yourself to trees, and make a damned nuisance of yourself?"

I paused. "My dad was a ranger and superintendent in the State and National Park system in America, and my mother was a fisheries biologist. I believe in conservation, but I don't plan on chaining myself to any trees."

The man considered my response. "O.K., Paul. I think we'll get along just fine. But just one more thing. Don't bring any locals into the compound."

"All of my friends are Samoan."

"You are welcome to have anyone you want as a friend. All I'm saying is that we don't want you to bring any Samoans into the compound."

I decided that I didn't need the accommodations after all.

Seventeen miles past the Taga blowholes, our van climbed the hill above Samata. As we rounded a corner, a huge truck filled with rain forest logs whizzed past us on its way to the wharf.

We passed through Falelima, Neiafu, and several other villages until we came to a small sign pointing to a little dirt road leading five miles through the forest to the sea. "Falealupo Village," the white letters proclaimed on the official green background. I was excited to see the village, but I was also fearful. What the Samoans regarded as the home of Nafanua and the entrance to the world of spirits, up until now had been only a point on a map. It would now become our home for the better part of a year.

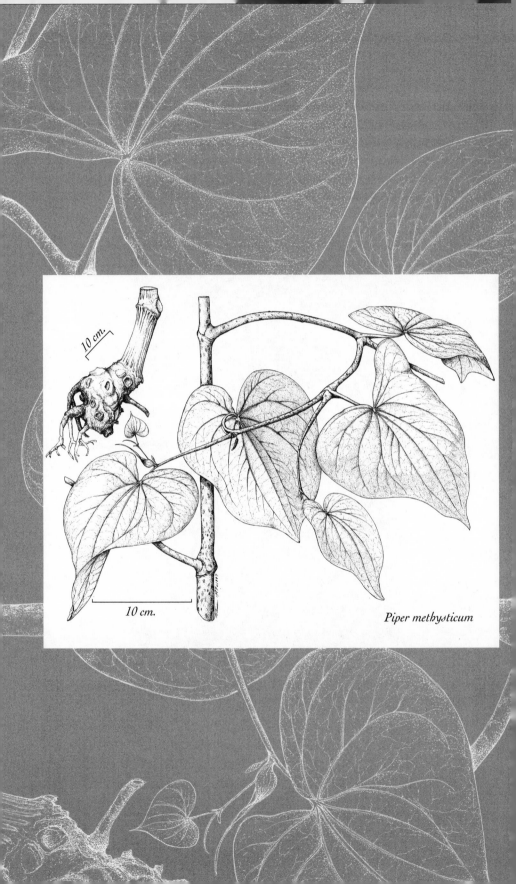

10 cm.

10 cm.

Piper methysticum

CHAPTER 2

Arrival

*Savaii. After the rain, when the
sun is shining and you walk
through the bush it is like a hot-
house, seething, humid, sultry,
breathless, and you have a
feeling that everything about
you, trees, shrubs, climbing
plants, is growing with an
impetuous violence.*

W. Somerset Maugham,
A Writer's Notebook

A s we turned down the small track toward Falealupo, it seemed as though we were nearing the edge of the world. But we were in fact less than a day's journey from the Mulifanua wharf, which is within sight of Western Samoa's international airport. While the distance from the United States, measured in miles, is great, that same distance measured in travel time is relatively trivial. Our return would require a three-hour drive back to Salelologa, with another three hours on the ferry back to Mulifanua. Western Samoa is about ten hours flying time from Los Angeles, so total travel time back home, not counting time spent waiting for the ferry, immigration, stopovers, and customs, would be only about sixteen hours.

Over the past several centuries there has been a huge reduction in travel time between Savaii and the West. Early explorers typically conducted coastal surveys or scientific investigations en route, making direct comparisons of travel time between then and now difficult. But the voyage of HMS *Pandora* to Samoa in 1790 is perhaps a standard of rapidity: the Admiralty ordered the *Pandora* to sail with all possible speed to the South Pacific in pursuit of the *Bounty* mutineers. Captain Edwards left England on 7 November 1790, spent a few weeks in Tahiti capturing fugitives, and continued west, sighting Savaii on 18 June 1791.[11] Total elapsed travel time: about six months.

This extraordinary reduction in the temporal distance of Savaii from six months to sixteen hours from the West had major implications for studies of natural history and ethnobotany. Even though I intended to stay with my family in Falealupo for extended periods, intermittent visits to New Zealand, Australia, and the United States were not only possible, but planned. My scientific predecessors of an earlier era had no such luxury. They were unlikely to have been able to make more than a single expedition to the South Pacific in their lifetimes. Botanists such as Joseph Banks of Captain Cook's *Endeavour,* Charles Darwin of Fitzroy's *Beagle,* or Charles Gaudichaud-Beaupre of Freycinet's *l'Uranie* depended on their notebooks and specimens from a single voyage for the rest of their scientific careers.

Our situation was far different. While resident in Falealupo I could depend on the system of ferries and international aircraft for evacuation of my family in case of injury or serious illness. I could be easily reprovisioned with scientific equipment, personally consult with overseas colleagues, and, most importantly, rapidly receive results of pharmacological testing of my plant samples.

But temporal proximity to the West carries disadvantages as well. Even though I had chosen Falealupo because of its geographical remoteness, the village could not long resist the gravitational pull of Western culture. The arrival of an ersatz Samoan culture from Apia and Pago Pago (the capital cities of Western and American Samoa) would herald an increased cultural erosion. Although relatively resilient, traditional Samoan culture is still vulnerable to incipient Westernization. Even my ethnobotanical research might accelerate the decay of Samoan culture, through a perverse anthropological variant of the Heisenberg uncertainty principle: the study of an object often changes its nature. I sought to reduce this likelihood by adapting our lives as much as possible to the Samoan way and by limiting my own impact on the culture. Robert Louis Stevenson and Margaret Mead became conspicuous *palagi* (foreigners) because of their writings and political involvements. Anthropologist Derek Freeman attained prominence in an Upolu village when he accepted a chief's title. But there seemed no reason why my ethnobotanical work should bring me to the attention of anyone beyond the circle of healers I would study with.

Temporal proximity also brings increased danger to the Samoan rain forest. The new wharves and the vehicular ferry we had just taken meant that even distant markets for timber or pulp could pose potential threats to the forest. And despite its grandeur, the Samoan rain forest is extraordinarily vulnerable to such threats. Evolutionary divergence in its plants and animals occurred after the initial biological pioneers colonized Samoa. As a result, the flora of Samoa has a greater percentage of endemic plant species than anywhere else in the world save Hawaii: nearly a third of the species of the Samoan rain forest are found nowhere else in the world.

The "splendid isolation," to use Darwin's phrase, of Samoa has also increased the vulnerability of the forest. Evolving in the absence of goats, sheep, and other grazing mammals, Samoan plants tend to lack the spines and thorns that protect continental plants. As a result, the Samoan rain forest is remarkably benign—it is possible to walk through it barefoot—but the forest is absolutely defenseless against attack.

It is modern humans in this age of diminishing remoteness who have posed the gravest threat to the Samoan ecosystem. Either possessed of sufficient wisdom, or simply deficient in destructive technology, the original Polynesian immigrants did not represent a significant threat to the rain forest. It is true that in the Falefa valley of Upolu one can see the remains of primitive waterworks, and there are places in Samoa where small-scale deforestation occurred. Perhaps one day an ornithologist will find, like David Steadman has in Tongan and Hawaiian caves, remains of bird species driven to extinction by the early Polynesians.[12] But in general, Samoa entered the twentieth century with its flora and fauna intact, a situation that continued until the 1970s, when timber prices skyrocketed and flying fox flesh attained value as a delicacy in Guam.

There were, however, neither logging trucks nor bat killers on the dusty little road leading to Falealupo. We descended a hill above a few taro plantations. To the side of the road two small children stood in front of a thatched *fale,* chewing on sugarcane. A man carrying two heavy baskets of green bananas suspended for balance on a stick across one shoulder stepped to the side so our vehicle could pass. Further down the road we passed several men weeding a taro patch. It was late afternoon, and the sweat glistened on their bodies. A swaybacked horse burdened with burlap bags of cut *Pandanus* leaves slowly trudged up the road, led by a small boy. We rounded a curve and descended another steep grade. The road seemed endless, leading us toward the tip of the Falealupo peninsula, the westernmost point of the Samoan archipelago where, by odd convergence of both Polynesian custom and the international date line, each day on this planet ends.

As we rounded another bend, the Falealupo rain forest came into view, a vast green canopy of leaf and liana, stretching three miles before disappearing in the haze. A few hundred feet more brought us to the edge of the forest, where the dirt road took on the aspect of a bower, framed on every side by huge trees. As we entered the forest, there was a perceptible drop in temperature. The cool air sparkled with the sounds of birds—not the exotic cacophony so often dubbed into Hollywood shots of "jungle," but the gentle purring of fruit pigeons calling to their mates and the rustle of honey-creeper wings.

Light filtered through the leafy canopy high overhead, giving the forest a play of shadow and iridescence like a Gothic cathedral's. Indeed, the forest light had a luminescent quality that could have been transmitted through a rose window. I thought of a Samoan couplet: *"Ua pa'ū le vao, ua liligo le taeao"* ("The forest echoes with sacredness, the morning is silent as dew").

All of us stared up at the long and twisted cables of woody lianas high overhead. Small shafts of bamboolike *Flagellaria* vines intermittently shot up toward the canopy. *Freycinetia* vines, with their long, linear leaves, covered the tree trunks. Beneath canopy and vine, the forest was open, consisting of a carpet of bright green ferns and an occasional sapling. The play of light and shadow, the massiveness of the tree trunks, and the openness of the forest floor reminded me of a redwood grove. Towering above the canopy, several massive banyan trees stood, their arms held aloft not by single massive trunks, but by a twisted maze of aerial roots. Their branches were covered with epiphytes: climbing ferns, twisted streams of *Piper* vines, *Collospermum* lilies with silvery swordlike leaves, and myriads of small, cherubic orchids, so rare that any one of them would incite a near riot among collectors.

I wanted to stop and walk through the rain forest, but the dictates of Samoan custom required me to first seek permission from the stewards and owners of the forest: the chiefs and orators of the village. We traveled several more miles through secondary forest and old coconut plantations until we reached a coastal lava flow, marked by a few stone cairns that appeared to be ancient tombs, and then passed a collapsed lava tube (a tunnel-like cave that develops as lava cools). Finally, we entered Falealupo village itself, consisting of brown thatched huts sprinkled along a mile of white sand beach.

A few breadfruit and frangipani trees provided contrast to the graceful silhouettes of coconut palms. Dugout canoes and a bonito boat with an intricately carved prow rested on the beach. A large fishing net hung on a rack to dry in the sun. Men with home-made goggles and spears were walking from the beach toward the thatched huts. Beyond the beach, the bright colors of coral heads appeared beneath the translucent blue waters. Falealupo, "House of the Alupo Fish," was clearly a fishing village.

Inland from the village we could see Faleū swamp, one of the last habitats of the endangered Samoan duck. A few women sat there beating their wash with sticks. In front of the swamp was a tidy white Protestant church. In the distance, rising from black lava rock, was a white Catholic church. A large mango tree and the small unpaved road separated the Catholic church from a deserted two-story house that had likely once served as the priest's residence. Not far from the church was a large *fale fono*—a chiefs' meeting house.

I stopped the car to ask a child the location of the *fale* of Pela Lilo. En route to Samoa, I chanced in Hawaii upon Pepe Burgess, a former acquaintance from Samoa. When Pepe discovered that we were headed for Falealupo, she urged me to contact her sister, Pela Lilo, who might have a small *fale* for rent. The child gestured toward a large oval *fale* with a high thatched roof inland on the white sand. After engaging the four-wheel drive mechanism, I drove across the deep sand to the house.

A family in a nearby *fale* spotted us and sent a child scurrying off to alert other family members. In a minute or two a young woman approached us, smiling. Our children clambered out of the vehicle, tired from the long ride. A stout woman appeared with a handmade whisk broom and proceeded to vigorously sweep the concrete floor of the *fale*. A man in a lavalava brought a large roll of mats and spread them for us in the *fale*.

We removed our shoes, entered the house, and sat cross-legged on the mats in front of the posts traditionally reserved for visitors. Soon a woman in her late sixties wearing an orange lavalava entered the hut from the other side. She had beautiful gray hair, soft skin, and brown eyes. Smiling, she walked over to greet us.

"Tālofa. I am Pela," she said.

I was struck by her grace and elegance, especially considering that we had just appeared unannounced in her *fale*. "My name is Koki and this is Barbara," I said. "We met your sister Pepe Burgess in Hawaii last month."

Pela kissed each of the children in turn, and laughed at Hillary's fascination with the Samoan broom. She smiled as a man who turned out to be her husband walked spryly with his cane into the hut. Though he appeared to be over eighty years old, he stood straight and tall. He was slender and his face had clean lines. Pela introduced him as Lilo. Since he was an orator, by custom his chief's title, "Lilo," had been adopted by his immediate family as their surname.

The children were still being introduced to the family when a procession of village chiefs entered carrying the traditional symbol of welcome, long kava roots. After they sat opposite us, Lilo, who knew I spoke Samoan, gestured for me to begin the formal Samoan rhetoric of welcome. I felt awed because I had just entered their *fale* as a complete stranger less than five minutes ago. But as instructed by Lilo, I began to recite the village *fa'alupega,* a string of honorific titles specific to Falealupo that I had memorized in preparation for our arrival:

Ia afio mai ma tala mai a'ao i le pa'ia o Auva'a ma aiga, ma le paia o le ma'opū o Nafanua.

Enter and welcome representatives of high chief Auva'a and his family, earthly emissaries of Nafanua.

Afio mai le Mātua o Lamositele ma Alo o Sina.

Welcome high chief Lamositele and the sons of Sina.

Ia maliu mai lava le mamalu o le to'afā ma Silia Laie, ma le mamalu o le tapua'iga.

Greetings and welcome to her four paramount orators and Silia Laie, and to those who assist them.

Pela smiled at my successful recitation of the *fa'alupega,* but the chiefs looked surprised. After a pause, one of the orators, an older man in a red lavalava and with a matching hibiscus flower behind his ear, cleared his throat and began the traditional speech of welcome for visitors:

"We meet in front of the village and not behind the village. We have the pigeon of our hope in hand, and have not lost the decoy. There are many words I could speak concerning the love of God, but they are insufficient. It is impossible to compare anything to the love of God because of its height, because of its depth, and because of its breadth, which covers the foundations of Samoa. The love of God surrounds us and flows unceasingly. When the untitled men have finished their preparations we will share the sacred kava drink. The sacredness of kava surrounds us and flows unceasingly."

The orator continued his traditional speech of welcome, and then looked at Lilo in expectation of a response. I slid forward on the mat, and carefully replied as I had been taught so many years ago:

"You have praised God this morning. We too are grateful for our safe arrival. As you have said, we meet in front of the village and not behind the village. We have the pigeon we have sought for, and have not lost the decoy. For this we are grateful. Concerning the kava: we are touched by this evidence of your love and respect. Even had you brought but one kava root, to us it would represent a thousand. Let us spiritually partake of it through the love of God."

"*Mālie,*" the chiefs said in approval. They motioned to the untitled men to put up the kava bowl. The kava roots were placed outside the hut.

I finished the traditional reply, introduced our family, and explained that I sought permission to study the plants used by the healers in Falealupo with the hope of discovering new medicines. I told them we hoped to spend a year in the village and pledged to respect their customs and to obey their chiefly authority. I also offered the use of my vehicle on Sundays to transport ministers of any denomination to perform their pastoral duties. At this, there were smiles of approval—apparently, as I expected, there were few operable vehicles in Falealupo.

One prominent orator introduced himself as Fuiono Senio. He was slender, with a strong build, and seemed to be the leader of the group. "We are astonished at your command of the Samoan language," he said, "and would like to listen to you speak more."

I smiled at his flattery, but realized the credit was due to Samoan chief Aumalosi, who had tutored me so many years ago in the ancient forms of Samoan rhetoric that are used with chiefs.

"Concerning your request, you are welcome to work in our rain forest and study with our healers. Pela is one of our best. We are also grateful for your offer of transportation for our ministers. The Catholic priest has to travel eight miles here to say mass, and the Congregational minister walks to the next village each week to hold services. I know they will appreciate your kindness."

Lilo then spoke. "This *fale* here," he said, gesturing with his hand to the hut in which we sat, "is now yours, for you and your family to use for as long as you want. We do not want you to pay rent—our family is now your family."

I turned to Barbara and quickly explained Lilo's greeting. Barbara nodded at me. Her earlier exposure to the language had allowed her to follow most of the conversation.

"You must be tired after such a long journey," Lilo continued. "Please make yourself at home. We have some food ready for you. You are welcome to join us as well," he said to the chiefs. The chiefs gracefully declined Lilo's hospitality, and left the *fale* after shaking our hands. "Come visit me," Fuiono whispered as he shook my hand. "I want to hear you speak Samoan again."

Pela's son Lamositele and her daughter-in-law, Fa'asaina, served us a simple meal of fish, pork, and boiled taro. A small tablecloth was spread for Barbara and the children. However, my taro and fish were served on a separate woven tray, a gesture of deference that made me uneasy.

After finishing our meal, Barbara and I took a stroll along the beach. Night fell quickly in Falealupo, and the children were tired after our long journey. Fa'asaina and Pela's granddaughters spread the sleeping mats in our *fale* and hung mosquito nets from cords tied to the intricately carved rafters. After prayers, we were soon asleep.

I woke shortly before dawn. I quietly removed the mosquito net that hung above our sleeping mat without waking Barbara and placed it on the shelf underneath the rafters. Mosquito nets still hanging after sunrise evidence laziness in Samoa. I walked barefoot across the white sand to the beach, and sat quietly awaiting the sunrise. Soon Barbara joined me, and we sat side by side as we remembered being in Samoa years before, when I was completing my doctoral research.

Breakfast was served by Pela's family on banana leaves as we sat cross-legged on the mats. The meal consisted of taro boiled in coconut milk, fried fish, bananas, and hot Samoan cocoa. The product of local cocoa beans that are alternately fermented and roasted, Samoan cocoa is a bittersweet substance that for most *palagi* is an acquired taste. The thought of drinking a hot beverage in the humid tropics is not appealing to most casual guests to Samoa, but what a world of taste and pleasure such fainthearted *palagi* miss. Our children enjoyed eating Samoan style with their hands. Fa'asaina brought in *apa fafano* (finger bowls) after the meal.

After breakfast, the children began their studies in a setting Barbara later affectionately referred to as "hut school." Before our departure for Samoa, Barbara had gone to each of the children's teachers, requesting textbooks, worksheets, and assignments for the next year. Sprawled on the mat, Mary played a math game on our solar-powered computer. Paul Matthew completed a social studies worksheet, while Emily read *Macbeth*. Hillary played next to the *fale* with some village children.

Midday brought a minor crisis. Although the swamp is used for washing clothes, the villagers rely on rain for drinking water. Our arrival in the village coincided with a period of drought. Fa'asaina

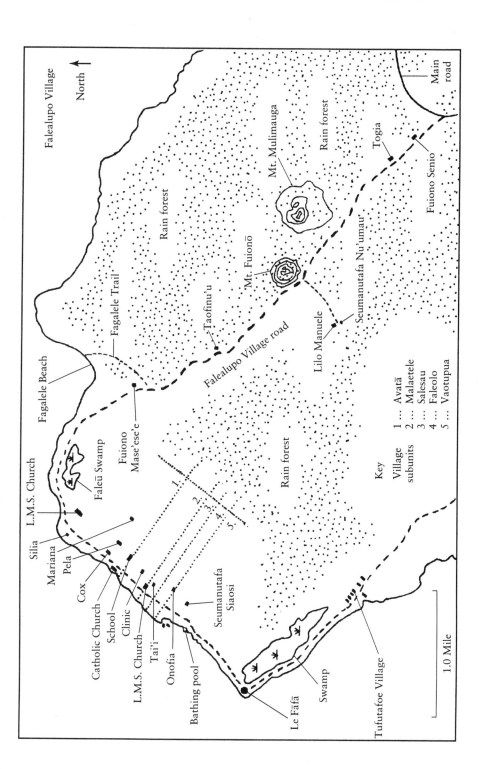

North

Falealupo Village

Rain forest

Main road

Mt. Mulimauga

Rain forest

Togia

Fuiono Senio

Fagalele Trail

Rain forest

Taofinu'u

Mt. Fuionō

Seumanutafa Nu'umau'

Falealupo Village road

Lilo Manuele

Fagalele Beach

Fuiono Mase'ese'e

Faleū Swamp

Rain forest

Key

Village   1 ... Avatā
subunits   2 ... Malaetele
            3 ... Salesau
            4 ... Faleolo
            5 ... Vaotupua

1
2
3
4
5

L.M.S. Church

Silia

Mariana

Pela

Cox

Catholic Church

School

Clinic

L.M.S. Church

Ta'i'i

Onofia

Bathing pool

Seumanutafa Siaosi

Le Fāfā

Swamp

Tufutafoe Village

1.0 Mile

carried into the *fale* a large bucket of water. As she sat the bucket down, I saw that the water had a brownish tinge.

"This tank is filled by rain gutters on the Catholic church," she explained. "We received special permission to use this water because of your arrival."

As *palagi* people we lack immunity to certain waterborne diseases that occur in Samoa, I explained to Fa'asaina. Would she be offended if we filtered our drinking water? We would be pleased to share the filtered water with her family.

Fa'asaina cheerfully agreed. We had come equipped with a Katydyn water filtration system manufactured in Switzerland. The system consisted of a hand pump surrounding a ceramic cylinder that screened out bacteria. Pumping water through the system proved to be an arduous task: it took ten minutes to filter just a bucketful of water.

After a light lunch, the children took naps, and then wandered about the village with their new Samoan friends. The return of the fishermen in their dugout canoes proved exciting for the children. They ran down to the beach to see the strings of colorful reef fish: parrotfish, angelfish, puffer fish, and some lobsters.

In the afternoon, I began my ethnobotanical research by interviewing Pela. The first interview is often the most important in an ethnobotanical study because it is crucial that rapport between the ethnobotanist and the healer be quickly established. The best ethnobotanists seek to situate themselves as apprentices or students of a knowledgeable indigenous person. Yet often a superficial asymmetry between the ethnobotanist and the indigenous person—an asymmetry reinforced by the scientist's access to economic resources, political power, or technology—must first be overcome. Assumed technological superiority is particularly important to eliminate, because possession of advanced technology can in some settings cause an ethnobotanist to be viewed more as a foreign necromancer than as a student. In field settings, it is thus important to demystify any imported technology as quickly as possible.

After chatting a few minutes with Pela, I wove into our conversation what I term an "informed consent script," which aims to obtain explicit consent for the use of each item of imported technology, beginning with the notebook and pencil and ending with all audio recording, photographic, video, and computer gear. Sitting next to Pela on a mat, I showed her my notebook and pencil.

"Pela, I lack the memorization ability that is so common among Samoans so I depend on some aids to help me remember our conversations. Do you mind if I take notes while we speak?"

She had no objections.

"And this is a tape recorder," I said, showing her my hand-held tape recorder. "Allow me to demonstrate."

I recited the *fa'alupega* of Falealupo village into the microphone. I then rewound the tape, and played back my voice for Pela to hear.

"Could I have your permission to use this to record our conversations?" I asked.

Pela seemed more apprehensive about the tape recorder than the notebook, but agreed. I explained to Pela that after my mother died of breast cancer, I wanted to make some contribution to medicine. I told her that it occurred to me that the Samoan people might know a lot about healing plants. Pela smiled.

"So I've come to Falealupo to learn about medicinal plants from the local healers. My hope is to collect the plants, take them back to the laboratory, and try to discover if they have molecules in them that might be useful as medicine. Would you be willing to help me?"

Pela paused a moment or two and then said, "We have many useful plants in the forest. I believe that God made each plant to be useful for something. I know a little bit about the medicinal plants and will teach you what I know. If a plant heals a Samoan, it will also heal an American or a Japanese person. Will you make pills from our plants?"

If I were successful in my research, I explained to Pela, a pharmaceutical firm might market a drug from the chemicals contained in the plants. There was only a very slight chance that I would discover a new drug, but if I did, I would protect her financial interests and the interests of the village.

"I'm not interested in money," Pela replied. "The plants are a gift of God. I never accept any payment for treating people. But if you make pills from any of our plants, I want you to bring some of the pills to me so I can use them to treat sick people."

Pela and I spent the rest of the day chatting in the *fale*. I was particularly interested in her concepts of disease. Pela told me that Samoans believe poor hygiene, bad diet, and interpersonal hostility to be major sources of disease. In the Samoan tradition, were there any other sources of disease?

"Some people say ghosts can cause illness, but I don't deal with that sort of thing," she replied.

We learned that Falealupo village is divided into five parts called *fuiala* stretching along one and a half miles of beach from north to south: Avatā, Malaetele, Salesau, Faleolo, and Vaotupua. Pela lived in Avatā, in the north. A half mile south in Vaotupua is a large pool called Salia that fills with brackish water at low tide. Since Falealupo has no running water, we were forced to walk or drive down to Vaotupua for our evening bath. As the first *palagi* family to have taken up residence in the village, we were very much a novelty. Our initial family ablution provided major entertainment for the village children. Twenty or thirty gathered while we bathed, watching carefully to see if one of our lavalava would immodestly fall off.

Our children were self-conscious bathing in front of a jury of their peers, so I sought a diversion. To entertain our children as well as the village children, I had brought to Samoa a "Mr. Magic's Balloon Sculpture Kit," consisting of a fish-shaped pump and a generous supply of long balloons. While Barbara and the children bathed, I called the village children away from the pool, and began to produce a variety of brightly colored balloon sculptures for them. With a bit of imagination, my balloon creations could be said to resemble giraffes, dachshunds, or Darth Vader light sabers. Even though none of the village children had heard of, let alone seen, giraffes, dachshunds, or light sabers, their enthusiasm for my productions was tremendous, causing a near riot as older children climbed over the younger ones to seize the balloons. Finally, with the bemused aid of a passing chief, I restored order to my little market scene. I thought myself quite clever as the children dispersed throughout the village shouting "Paluni (Pah-loo-knee)! Paluni!" ("Balloons! Balloons!"), their attention successfully diverted from our family bathtime.

On our next visit to the bathing pool, we found a crowd of fifty children waiting, including, I believe, some children from neighboring villages, all yelling with anticipation "Paluni! Paluni!" At that moment, our dream of private family bathing burst. "Mr. Magic's Balloon Sculpture Kit" made its final appearance that night.

After bathing, we gathered with Pela's children and grandchildren in our *fale*. While we sat cross-legged on the mats, Lamositele began pumping the Coleman lantern. Lamositele's hands were strong, with protruding veins. His lined face gave him a regal bearing, but one tinged with sadness. Soon the *fale* was filled with the lantern's warm glow and comforting hissing sound. Lamositele pulled out a large red Catholic Bible and read a chapter of Psalms aloud to us. His voice was rich and filled with devotion. As he began to pray, his voice became stronger, typifying the deep faith of the Samoan people.

I have never heard anyone pray like the Samoans. Samoan prayers are long, lasting up to twenty minutes, but also differ in many other ways from prayers uttered by Europeans. First, Samoan prayers are completely extemporaneous, regardless of the person's church or religious affiliation. Second, Samoan prayers are family prayers. At prayer time, all activity in the village ceases as each family unites in singing and praying. In Falealupo, which lacks electric-

ity, motors, and other industrial noise, perhaps fifty or more different families could be heard at prayer time singing praises to God.
Third, Samoan prayers represent personal struggles with a God
who has the power to intervene in daily life. Lamositele addressed
God with the same language of respect used to address village
chiefs. To Lamositele, God is a personal being who watches over his
tiny island nation just as He notes the sparrow's fall. During World
War II, entire villages prayed nightly that Japanese invaders would
never come to their shores. Unlike many island countries, Samoa
was spared the horror of war. Samoans seek blessings for their crops
and fishing, and their harvests are usually abundant. But Samoans
do not always have their prayers answered affirmatively, as Lamositele knows. Despite prayers, some of the sick die, and during hurricane season, Samoa too is visited by violent storms.

As Lamositele's prayer drew to a close, the Samoan children
recited the Lord's Prayer and then closed with a hymn. Although
they did not understand Samoan, our children were startled by this,
the first Catholic recitation they had heard. I was deeply touched by
Lamositele's willingness to include us in the evening services. Gallup
polls show that most Americans pray, but few other than evangelics
would consider involving strangers in this confidential aspect of
their personal lives.

Although the deep religiosity of the Samoan people has sometimes been ignored by students of the culture, it plays an integral
role in Samoan life and in the Samoan world view. In an earlier age,
European observers were not as reticent about Samoan religious
practices as they are today. In 1862, the crew of HMS _Fawn_ spent
their first night on Ta'ū island in the extreme eastern part of the
archipelago. As they played card games in a Samoan _fale,_ they
became aware of Samoan vespers. "After dinner, the people seemed
much amused by watching us play at whist and backgammon,"
T. H. Hood wrote.

> Our amusements were suddenly suspended by hearing the
> master of the house singing the evening hymn, in which most
> of the assembly, taking their books out of their waist-cloths,
> joined; and we, removing our hats also, added our voices to
> the best of our ability, hymn-books being handed to us. The
> psalm finished, our host made a long extempore prayer;
> the Lord's prayer concluding the service. It certainly was a
> striking scene, half a dozen unarmed Englishmen sitting here
> in the midst of a crowd of half-naked islanders, and
> receiving a lesson of this kind from people so recently
> designated 'ferocious savages.'[13]

After dinner, I read to the children from *James and the Giant Peach*. We rolled out the sleeping mats and tucked the edges of the mosquito nets around the children to avoid encounters with centipedes. Behind our *fale*, in the cook hut, a few village men gathered and invited me to share kava after Barbara and the children were asleep. The bitter, slightly numbing beverage seemed to relax everyone. Their conversation concerned the projected night's fishing expedition. Even though I was a foreigner, they made me feel welcome.

The informal gathering broke up as the men went to strap kerosene lanterns onto the prows of their canoes. Each dugout had a kerosene lantern strapped to its prow with coconut cord and a white tin plate positioned to reflect light downward for spearfishing. The lighted canoes swirled about the dark lagoon like fireflies. I soon fell asleep in my mosquito net, with visions of lighted canoes moving out to sea.

The next morning, I continued my interview with Pela by discussing Samoan disease categories.

"What are some examples of diseases you treat with plants?" I asked.

"That vine over there," Pela said pointing to a beach morning glory, *Ipomoea pes-caprae*, "is useful for *mūmū*."

"That refers to something that is red and swollen, doesn't it?"

"Sometimes," Pela replied.

"What else does it refer to?" I asked.

"There is *mūmū ta'ai*," she said. "It's associated with fever that goes around in the body but can't come out. And there's *mūmū fefete*. The skin swells, but there is a palpitation associated with the swelling."

When Pela finished, I studied the list of fifteen different varieties of *mūmū* that I had written in my notebook, ranging from partial paralysis to skin eruptions.

"These seem to be such different symptoms," I said. "What unites them all into a single disease?"

"I don't know," said Pela. "I just know that they are all types of *mūmū*."

"How could I tell if I saw someone with *mūmū*?" I asked. "Is there any single indication that would tell me that someone has *mūmū* instead of some other disease?"

"You just have to learn them," Pela said.

Pela and I left our *fale* and strolled along the beach. Picking up a beach morning glory, Pela said, "We call this plant *fue moa*. Sometimes the children play with the fruits like little tops. But it is very useful for _mūmū_. And that plant," she said, gesturing toward a beach pea, _Vigna marina,_ "we call _fuefue sina._ I use it to treat _mūmū lele,_ a condition that sometimes proves fatal to new mothers."

I collected samples of both plants and returned to our _fale_ with Pela. I attached the solar panel to the computer and recorded the data for each collection: the plant's Latin name, the precise location where I collected it, the date of collection, and the local names and medicinal uses Pela had described. I next spread a piece of the plant on a sheet of newsprint, wrote its Latin name and collection number on top of the sheet, and placed the sheet inside blotters in the plant press. Pharmacological specimens followed. On a cutting block I chopped up the remaining parts of each plant and stuffed the pieces into Sigg aluminum bottles, originally manufactured in Switzerland for transport of fuel by mountain climbers. I poured ethyl alcohol into each bottle to preserve the plant, slipped a small piece of paper with the plant name and collection number written in pencil into the top of the bottle, and labeled the outside of the bottle with the collection number and plant name with a marking pen.

During our first week in Falealupo my work with Pela progressed, but not as I quickly as I had hoped. To get a better idea of Pela's concepts of diseases, I asked her to name every disease she knew as I wrote each name on a three-by-five card. Soon we had a large pile of cards on the mat in front of us. I then handed the cards to Pela and asked her to place them in groups.

"How do you want me to group them?"

"However you wish."

Pela fiddled with the cards, and then began to place them in piles. One pile consisted of all of the _mūmū_ diseases. Another pile consisted of _ila_ diseases, skin ailments she identified as occurring in children. Another stack of cards was for _tulitā_—a group of diseases characterized by abdominal distress. Soon there were several piles. It appeared that in Pela's system most diseases had two names—a generic term such as _mūmū_ and a specific modifier such as _lele_ combining to form _mūmū lele._ Pela's card sorting was entirely consistent with a theory developed by anthropologist Brent Berlin, who argued that indigenous peoples usually adopt binomials in classifying the natural world. Based on studies in Central America, Berlin also suggested that indigenous peoples invariably employ five levels of hierarchy in their classification schemes. Pela had documented the first two. Excited to see the next hierarchical level in Samoan classification, I produced string and Scotch tape.

"What I'd like you to do, Pela, is to show me how the diseases in these different piles are related to each other."

Pela looked puzzled.

"For example, if these two piles are closer to each other than they are to any of the other piles, I'd connect them with string like this."

I stretched the string between the two piles and taped the ends onto the top cards.

Pela looked frustrated, but seeing my eagerness, she pointed out several different piles of cards to connect with string. After fastening the string between the piles with tape, I asked her what she called these new groupings. Pela was again puzzled.

"For example, Pela, you've connected the *mūmū* pile and the *ila* pile. Why?"

"Because you wanted me to."

"Yes, but what do these piles have in common with each other?"

"I don't know."

I explained to Pela that many people group birds or plants into species, genera, and families. Different peoples have different types of classification schemes, with different types of hierarchy.

Pela looked at me. "I really don't think about things that way," she said.

"So this pile here that you call *mūmū* doesn't have any relationship to any other group of diseases?"

Pela shook her head. It was apparent that although she used binomials to categorize diseases, she had no higher level of taxonomic hierarchy, a remarkable finding. Samoan disease classification consists of unrelated groups of binomials. Some of the words Pela used, such as *ila* or *mūmū,* appear in disease classifications in Tahiti and other distant islands. Perhaps anciently there was a well-articulated Polynesian disease classification system, with consistent connections between different modules.

But I was disturbed by the apparent arbitrariness of the task I had set for her—Pela had begun to tape connections between the piles merely to appease me. Like Procrustes, who guaranteed a perfect fit to his bed for all guests by either chopping off their legs or stretching them on a rack, I had attempted to force Pela's knowledge into divisions of my own making. Fortunately Pela resisted. Had she complied, I would have recorded not the healer's perception, but only an echo of my own preconceptions.

Our family life began to assume a certain predictable but enjoyable routine very soon in our stay. Because of her old age, Pela could work with me for only an hour at a time, once or twice a day, before she tired. I spent the rest of my day collecting and preparing plants, exploring the forest, and chatting with other villagers. Interjected throughout the day would be numerous unannounced visits of villagers to our *fale,* each usually bearing a gift of food, be it a stalk of bananas, a piece of pork, or some lobster. None ever asked us for anything; the sole purpose of their visits was merely to take joy in one another's company. As I return from Samoa to the United States it is these little unannounced visits, so different from modern American ways, that I miss the most.

This ongoing pattern of reciprocal visits gave village life a feeling of civility and elegance that often is lacking in technologically advanced societies. Work and pleasure were integrated in a seamless fashion, with friendships renewed on a daily basis. The thatched huts without walls allowed little privacy, but facilitated social interaction, as it was easy to see if someone was at home for a visit. There was no door to knock on or bell to ring; if one wished to visit a neighbor, one merely walked into their *fale* from any direction and sat down on the woven mats. Robert Frost ironically suggested that "good fences" do not good neighbors make; after living in Falealupo I wonder if walls should be reconsidered as well.

Of course, the absence of walls gave a different meaning to privacy, which, strictly speaking in the Samoan sense, ended with the boundary of one's lavalava or other clothing. All of us quickly realized that in such a situation a higher standard of behavior is required. In Falealupo there is no such thing as a private quarrel or a secret melancholy. All of one's life, from morning to night, is open to public inspection.

The flip side of the extraordinary openness of Samoan life is the difficulty of discouraging intrusiveness. There is no Samoan word for "snoopiness," as two American schoolteachers near Apia, the capital city of Samoa, discovered. Even though this married couple resided in an European-style house rather than in a thatched *fale,* their frequent marital spats began to create significant interest among the neighbors. One evening before we went to Falealupo I was walking toward the home of the schoolteachers. I was startled to see a crowd of Samoan men, women, and children sitting silently in the dark next to an open window. Recognizing a Samoan friend among the dark forms, I approached him.

"What are all of you doing out here?" I asked.

"Listening," he quietly replied.

Suddenly there was an angry outburst from inside the house, followed by the sounds of the American couple's bitter quarrel.

"Why did you come here tonight?" I asked my friend.

"We come here every night," the Samoan said.

"How long have you been doing this?" I queried.

"About a month."

My *palagi* sensibilities scandalized by this invasion of marital privacy, I stood up to leave. In my haste I upset a large pot of Samoan cocoa and some ceramic cups carefully arranged on a mat. No Samoan entertainment is complete without refreshments.

But the loss of privacy in a Samoan *fale* is compensated by a constant connection to nature as well as community. It is said that the Japanese house is designed to give the illusion that one is within nature; the Samoan *fale* delivers the reality. From within our *fale* we had a constant 360-degree view of the environment. And frequently that environment intruded. To the delight of our children, there were plentiful and occasionally dramatic entrances by lizards, pigs, chickens, or sand crabs into our house. I saw to it that most of these uninvited visitors quickly exited with far less dignity, but sometimes it was delightful to lie with the children on the mats and watch a sand crab scurry through the *fale,* or listen to the chirping of geckos climbing the rafters. Hillary, our youngest, was particularly partial to small crabs.

Not all animal visitors were equally welcome. Rib-rough Samoan dogs, missing both fat and patches of hair, haunted the periphery of our *fale* during mealtimes. And we tried very hard to discourage the visits of a certain mangy kitten that would have supplied enough ecto- and endoparasites to equip an entire university parasitology lab.

The lack of privacy also affected my relationship with Barbara. To Samoans, public display of affection, even kissing or hand holding between married couples, is considered to demonstrate bad taste. We discovered that any semblance of seclusion in an open *fale* was possible only in the wee hours of the morning. Even then such seclusion is likely more imagined than real. Yet, with indomitable spirit during our first week in Falealupo, I invited Barbara for a private moonlit stroll along the beach.

"What for?" she naively asked.

"I thought it might be sort of romantic," I answered.

Barbara made a face and rolled her eyes. I was crushed. But she came to regret her impish impulse when, on subsequent occasions, she importuned me to accompany *her* to the beach at night. In a mock English schoolboy accent I replied, "Well, it wouldn't be considered *proper,* now, would it?"

But we both soon discovered that every evening in Falealupo is a rendezvous with loveliness. The dark silhouettes of the coconut trees, gracefully framed above by white banner clouds that, implausibly, had turned completely pink, and below by the absolutely white sand of the beach, compensated for many creature discomforts. Since Falealupo lacks electrical generators, cars, televisions, and other similar appliances, the transition from sunset to dark always proceeds with tranquility.

On Tuesday and Thursday nights the Catholic choir gathered in a *fale* near ours to practice hymns. Their a capella harmonies wafted through the village, turning it, for a moment, into a kind of immense outdoor cathedral. Thousands of miles from any source of air pollution, the stars burned brightly. And the peaceful culture of the Samoans made it possible to walk about the village at night without fear. During our evening strolls, after choir practice, we often encountered young people quietly strumming a guitar and singing an island song underneath a tamarind tree.

While lush in sight and sound, Samoa, for us, was not a copy of a Disney-like paradise, but was populated with real people whom we began to care deeply about. And we found that our love and concern was soon reciprocated. "Island life," an ill, but productive, visiting writer told Samoan resident H. J. Moors in December 1889, "has charms not to be found elsewhere. I like this place better than any I have seen in the Pacific."[14] So it was with Robert Louis Stevenson, and so it was with us.

1 cm.

10 cm.

1 cm.

Syzygium malaccense

CHAPTER **3**

Missionary

This was the first time, in all my years in the Pacific, I had ever exchanged two words with any missionary, let alone asked one for a favour. I didn't like the lot, no trader does; they look down upon us and make no concealment; and, besides, they're partly Kanakaised, and suck up with natives instead of with other white men like themselves.

Robert Louis Stevenson,
"The Beach at Falesa"

O n our first Saturday night in Falealupo, the village welcomed a new Catholic lay minister. Falealupo was intermittently visited by a traveling priest, but it had been some time since a Catholic cleric had been in residence in the village. Since the majority of the villagers are Catholic, the arrival of the new pastor promised to be a grand occasion. The minister arrived accompanied by over forty people, including his wife, family members, chiefs from his former village, and other well-wishers. All of the Falealupo chiefs assembled in a *palagi*-style house ornamented with flowers and palm fronds to greet the visiting party. After everyone was present, the Falealupo chiefs began a *fa'atau*, or ritualized debate, to determine who would offer the speech of welcome on behalf of the village. The contest is largely for show, since the outcome is seldom in question—in Falealupo, one of the four paramount orators (Fuiono, Ta'i'i, Soifuā, or Taofinu'u) is invariably selected to represent the village on such occasions.

Sitting cross-legged on a mat next to Fuiono Senio, I listened as each of the chiefs was ceremonially asked if he would like to give the village speech. Each respectfully declined. I was surprised when the *fa'atau* turned to me, but I courteously refused the opportunity. Fuiono leaned over to me and whispered, "You don't understand—we want you to give the speech."

After some chiefs were asked again, I was again ceremonially queried, and this time accepted. I recited the *fa'alupega* of both villages and offered, using proverbial expressions, the welcome of Falealupo to the new minister. At the conclusion of the speech, the visiting chiefs burst into applause, clearly surprised by the novelty of a *palagi* speaking chiefly language.

Woven trays filled with hot breadfruit, *palusami* (young taro leaves baked in coconut cream), and baked bonito were placed in front of each person at the gathering. Over dinner, I engaged in a bit of banter with the new Catholic lay minister.

"I understand that celibacy is important for Catholic ministers, yet I see that you have come with your lovely wife. Has new doctrine been announced by the Pope?"

Without skipping a beat, the minister replied, "Indeed—I have wives in several villages."

There was laughter among the visitors.

"In fact," he continued, "I have come with a new wife for you. Meet Iolani," he said, gesturing toward his three-hundred-pound maiden sister.

"I'm sure Iolani is wonderful. But I am already married and have a wife."

"Yes, but there is nothing wrong with taking another wife. The Bible records Abraham and later David and Solomon accepting multiple wives."

"For me there is a more important authority than Scripture to consider. You don't know Barbara."

"What? You are afraid of your wife?" he asked.

"Yes," I said. "*E laititi ae maigi:* She is small but dangerous."

The crowd convulsed in merriment, and the dancing began.

The next day was Sunday. My intention was not to do any research on Sunday, but to spend the day with my family and attend church. Some of my scientific colleagues have gently asked whether my devotion to Christianity is compatible with advocacy of indigenous rights and preservation of indigenous cultures. Given the checkered record of missionary interactions with indigenous people, particularly in Polynesia, it is a perfectly reasonable question, and I try to respond with candor: I am committed to my faith. In my ethnobotanical work I do not seek to preach my faith to indigenous people, but I believe that my religious commitment facilitates a more empathetic response to indigenous approaches to the divine.

For me, sacred ground remains sacred ground whether revered by a Christian, a Buddhist, a Sikh, or an indigenous Animist.

I arose at dawn to prepare for our church services and drive the ministers of other denominations to theirs. After depositing the new Catholic lay minister at his church inland near the main road, I drove four miles further to Sataua village, where I picked up Mormon branch president Tu'imauga Alavine and his wife. Together we returned to Falealupo to Onofia's family *fale,* which serves on Sundays as our chapel. I then picked up Filiva'a, the Congregational minister, at his home and drove him two miles along the beach road through coconut groves past 'O le Fāfā (the entrance to Pulotu) and on to Tufutafoe, where he conducted Congregational services for that village. When I returned to Falealupo, the road was filled with several hundred Samoans: women in white dresses and large hats, men in white shirts and lavalava, and small children bedecked with flowers, all walking to the Falealupo Congregational Church, where the deacon conducted services. At the conclusion of our own Mormon service in the Vaotupua section of Falealupo, I returned all of the various ministers to their residences, finally arriving at our *fale* for *to'ona'i,* or Sunday dinner, at 1:00 P.M.

Although mass had concluded far earlier, Pela, Lilo, and Lamositele had delayed their Sunday dinner for me. Soon the *to'ona'i* was laid before us in all its glory. In remote villages, *to'ona'i* is more than a celebration—it is something of a nutritional necessity. Busy in subsistence agriculture and reef foraging, active individuals can drop into a caloric deficit during the week, but they make up for it with Sunday *to'ona'i,* consuming on average, as Joel Hanna at the University of Hawaii discovered, a whopping 5,300 calories.[15] Since men do most of the cooking, providing a superb *to'ona'i* becomes an issue of masculine pride. Preparation for the meal begins on Saturday night when the canoes are launched for the midnight fishing expeditions. Long before dawn, young men light the fires to heat the rocks that will be piled upon the leaf-wrapped packets of food to create an *umu,* or Samoan stone oven. By Sunday morning, smoke pervades the village. Hot baked taro, breadfruit, *palusami,* coconut cream baked in coconut shells, baked reef fish, *oka* (raw marinated fish), vermicelli noodles simmered in soy sauce, corned beef, and lobsters are all provided in abundance. Our first *to'ona'i* in Falealupo was superb, and brought back for me many memories from years past.

After dinner, Lilo asked how I had learned to speak chiefly Samoan. Over a cup of raw cocoa, I explained that at the age of nineteen, I was called by the Church of Jesus Christ of Latter-Day Saints to perform two years of unpaid missionary service in Samoa.

On arrival in Apia, I was immediately dispatched to Savaii on a copra boat. My pastoral duties were simple: with my companion, I was to look after our few members, do whatever I could do to assist the local people, and explain to anyone who wished to listen the message of Christ's gospel.

The seas were rough on that first Samoan voyage. Aboard there were no chairs, benches, cabins, or even free deck space. For safety, there was only a single dilapidated throw ring to service seventy people. The one-line fillers that appear in newspapers came to mind: "Ferry capsizes in Bangladesh, 350 perish." I finally perched precariously on a bag of copra next to a Samoan woman, who was silent for the entire trip. This was just as well, since my Samoan language skills were limited to a few greetings.

I tried my best not to become seasick, even though the combination of high seas and diesel fumes made it difficult. After several hours, the waves began to calm, and I saw that we were approaching the dark land mass of Savaii. The rain forest grew right up to the edge of the sea along that coast: a palisade of large trees, perhaps a hundred feet tall, whose brilliant green crowns contrasted with the black lava rocks that broke the waves. As far as I could see toward the distant mountains, the rain forest covered the landscape.

It took five hours through dust and potholes to reach Safune village. Our vehicle stopped in front of a small *fale* on the beach that would be my home for the next six months. Inside was a woven sleeping mat on which someone had placed a pillow and a carefully ironed hand-embroidered pillowcase. Next to the hut was a tiny open-air chapel made of several sheets of corrugated tin held aloft by crude wooden poles. The floor was crushed coral.

I felt like a page ripped from a romantic novel. Not only was I one of the few *palagi* in Savaii, I was also one of the few Mormons. The next day I met High Chief Aumalosi. I didn't have the slightest idea what he was saying to me. Rather than becoming frustrated, Aumalosi sat on a mat in front of me, began a sentence in Samoan, and motioned with his hands for me to repeat it. I tried my best to mimic his sounds. He spoke again, motioning for me to try again. Over and over for the next hour Aumalosi repeated the same sequence of unintelligible syllables until I finally reproduced them to his satisfaction. He smiled and stopped. Our first lesson was over. Several days a week for the next six months, Aumalosi returned. Slowly I memorized long sequences of chiefly Samoan. Each time Aumalosi patiently added a few words to my repertoire until I could recite long sequences of strange sounds. Eventually I began to understand the texts that he helped me to memorize: ancient genealogies, key proverbs, fragments of kava speeches.

I still do not know why I was chosen for such linguistic training. Chiefly rhetoric is almost never taught to a *palagi*, not because
the skill is hidden from outsiders, but because fluency in common
Samoan is required—Seuss before Shakespeare. Fortunately, this
prerequisite was waived for me.

The cocoa was nearly finished when I concluded my explanation to Lilo and the few other chiefs who had gathered in the *fale*.
Lilo assured me that Aumalosi had taught me well. As the cocoa
pot was taken back to the cook hut for refilling, the conversation
turned to fishing. I asked about the use of plants in fishing, particularly as fish poisons. For centuries, Polynesians have used the roots
of *Tephrosia piscatoria*, a vine that grows along the beach, for fishing. They place the macerated roots under a coral head. Within a
few minutes, reef fish surface belly-up.

In response to my query, a chief related an extraordinary occur-
rence from the Safata district on Upolu. Apparently several years
before my return to Savaii, a foreign consultant had advised the
Samoan government that the traditional plants used to poison fish
should not be used anymore because they posed a grave peril to the
reef community. The government dutifully accepted the recommendations and banned all uses of fish poisons.

Ever resourceful, the people of Safata district seized on dynamite as a potent substitute for the traditional plants, discovering
that explosive shock waves stun fish just as effectively as plant poisons. One day a sixteen-year-old boy was given his first opportunity
to throw the dynamite to the coral head. But as soon as the fuse was
lit, he was paralyzed by fear. Despite the screamed warnings of the
other villagers, he just stood there as the fuse burnt down.

The boy was some distance from the rest of the fishing party, so
the detonation did not seriously injure anyone else. The boy himself appeared to be dead. Several shaken villagers cradled his body
and placed it in the back of a pickup truck for an emergency run to
Moto'otua Hospital. They covered him with burlap bags and leapt
into the cab. Just then someone suggested that since they were going
to Apia, perhaps they should throw in some bags of taro to sell at
the market after they visited the hospital. Several large burlap bags
already filled with taro were quickly loaded into the truck next to
the body. The truck then started for Apia. At the outskirts of the village, Chief Leofo Salemalama waved the truck down.

"Where are you going?" he demanded.

"To Apia," came the breathless reply. "Jump in!"

Leofo clambered into the back of the truck, where, nestled
alone among the bags, he had a unique opportunity. Traditionally,
a chief's status is reflected in the quality of the taro or *Dioscorea*

yams he grows. Possession of advance intelligence concerning the quality of the crops produced by neighboring villagers allows, if necessary, for adjustment of one's agricultural effort and strategy. And so, unbeknownst to the frantic driver and other occupants of the cab, Leofo began, as the truck neared Apia, to surreptitiously peek at the contents of the burlap bags.

The first bag contained blue taro of medium size. Leofo was pleased; he knew that his blue taro was better. The second bag contained very large *talo palagi,* a white cultivar, which impressed Leofo, but he was still confident that his upland plantation could yield even bigger taro. But when he lifted the next burlap bag, he spotted the mangled corpse of the dynamited boy.

"Quick! Quick! We have a dying person in the back!" the driver yelled as the pickup screeched to a halt at the hospital. Running to the truck, the orderlies quickly loaded a body on a stretcher. But as they ran with the stretcher into the hospital, the driver saw that the litter carried not the victim of the explosion, but the unconscious Leofo.

"No! Not him!" the driver yelled. "The other dead guy!"

At this line in the story, I began to laugh, but the Samoans in the *fale* were not amused. "What's so funny about that?" they demanded.

Squelching my mirth, I explained that the mix-up with poor Leofo struck me as amusing, though of course the dynamite death was tragic. Looks of misunderstanding filled the *fale.* "Why do you think Leofo's situation was funny?" they asked incredulously. "He's lucky he didn't die from the shock."

Sensing a cultural dissonance, I pursued this last statement. "I'm sure that Leofo was tremendously surprised, even shocked. But why do you say he was in danger of dying?"

"Don't you understand?" one villager said. "Sudden surprises or shocks like that can kill!"

"You mean Leofo might have had a heart attack?" I asked.

"No, no. You don't understand. When one has a shock like that, their *toala* can jump. And if your *toala* moves too quickly, or in the wrong direction, it will kill you."

"Where is this *toala?*" I asked.

A chief pointed to a region beneath his navel.

"And what happens if it moves, but not far enough to kill you?" I continued.

"You get very sick. You have to seek out a healing woman. They know of plants and a way to massage the *toala* back to its normal place."

The next morning, I asked Pela about the *toala*. She explained that *toala* displacement does indeed cause illness and illustrated the type of massage that is used to treat it.

"How much would a patient pay you for this treatment?" I asked.

"*O lo'u fōfō o lo'u tofi lea mai le Atua:* My healing knowledge is my call from God," she said. "The plants are a gift from God. How can I accept money for what God has given?"

I spent the rest of the morning filling the plant press with specimens from the forest by the swamp, and then prepared for what I hoped would be a quick trip back on the ferry to Apia. I was running low on ethanol for the pharmaceutical collections, and was hopeful that I could obtain a significant quantity from the Apia pharmacy. I also needed to greet an important visitor at the airport.

The voyage back on the ferry was especially long and hot. I was grateful late that evening to drive into Potoga, the inland plantation of our Samoan friends and adopted family Dan and Kathy Betham, who insisted that we stay with them whenever we visited Upolu. The next morning, I was at the Apia pharmacy soon after it opened, lugging a jerry can I had purchased from the agricultural store. The pharmacist's eyes grew large as he learned what I wanted, but after I explained my research, he graciously consented and carefully poured twenty-five liters of 95 percent ethanol into the can. Mission accomplished, I went to the airport to await my visitor—a physician to help me with my research.

Bringing a physician to learn from indigenous healers bucks nearly two centuries of well-established colonialist tradition. "Medicine was taken as a prime exemplar of the constructive and beneficial effects of European rule," writes historian David Arnold, "and thus, to the imperial mind, as one of its most indisputable claims to legitimacy."[16] Since Western medicine was regarded as prima facie evidence of the intellectual and cultural superiority of Europeans, it is not surprising that indigenous medicine was denigrated by medical practitioners. In her study of colonial medicine in Africa, Stanford researcher Megan Vaughan found that "traditional African culture and society was constituted as the major enemy of social and medical progress, and was often represented by the figure of the 'witch doctor.'"[17] Out of the European exploration of Africa was born an entire genre of popular literature, the jungle doctor books, in which a white physician, fighting the forces of ignorance and superstition, brings modern medicine to a benighted people, often in the process engaging in a confrontation with the local medicine man. The final chapter of a representative volume published in 1952, *African Jungle Doctor: Ten Years in Liberia*, is even entitled "The Defeat of the Medicine Man." The protagonist, Dr. Werner Junge, "successfully replaces a nose with a plastic substitute and cures a case of hysterical paralysis through hypnotism," Vaughan reports. "After these 'triumphs' the community apparently pronounced that the local 'medicine man' was 'old-fashioned' and he shut up shop." Junge's response to these events:

> This was precisely what I had been striving for. It was for them to see for themselves that their ideas of sickness and magic and enchantment were old-fashioned. The tyranny which the primeval forest and the black medicine man exercised over their souls and bodies had been broken.[18]

Denigration of indigenous medicine was of course not limited to Africa. "The Samoans in their heathenism," wrote early missionary George Turner in 1884,

> seldom had recourse to any internal remedy except an emetic, which they used after eating a poisonous fish. Sometimes juices from the bush were tried; at other times the patient drank on at water until it was rejected; and, on some occasions, mud, and even the most unmentionable filth, was mixed up and taken as an emetic draught.[19]

Turner protested the Samoans' use of "unmentionable filth" because he had an alternative: he had begun to manufacture and dispense to the Samoans his own medicinal powders and preparations.

> Whether I would or not, I was obliged to turn out "Graham's Domestic Medicine" and become *head doctor* of the district. Day after day I had twenty, thirty, or fifty calls for advice and medicine. I appointed an hour, morning and afternoon, for the purpose, and, by making a small charge of something useful to the servants, such as a hank of cinet [sinnet—cord woven from coconut fiber], or a few taro roots, for a dose of medicine, I was able to keep the rush and inconvenience within bounds.

Thomas Trood, son of the British Consul in Apia, reported that by 1866, the practice of pharmacy by missionaries had become widespread:

> In those days the missionaries, as far as they knew, supplied the public with medical comforts and medical advice. . . . The missionaries gave out liberally many other medicines

besides salts, properly insisting on the natives making a fit return in fowls and vegetables, & etc., but not cash.[20]

Some visitors criticized such indiscriminate provision of pharmaceuticals, however. "The missionaries dispense medicine to their people," wrote A. B. Steinberger, an American emissary to Samoa in 1876.

> This is a grave error. Excepting Dr. Turner, of Apia, none are regularly-trained physicians. They adhere to the old school of practice, and ignorantly dispense blue-mass, gray powders, calomel, and other preparations of mercury, while Dover's powders, podophyllum, preparations of arsenic, & etc. are freely given. I foresee in this reckless issuance of drugs no little mischief in the future, as mercurial diseases must certainly develop themselves unless it is abandoned.[21]

Steinberger's report raises an interesting question: if one were sick in Samoa in the nineteenth century, would it be preferable to be attended by a European physician, who bled his patients and dosed them with mercury and arsenic preparations, or to be attended by a Samoan healer, who massaged them and dosed them with herbal preparations? Apparently this question was not a hypothetical one by the beginning of the twentieth century. Lieutenant Commander Hunt of the U.S. Navy Medical Corps in Pago Pago in American Samoa disclosed that by 1923 many of the *palagi* residents of Samoa, including missionaries, frequently resorted to Samoan healers. "A majority of the Samoans believe that their own crude drugs and harsh medical treatment are more efficacious," Hunt wrote with some astonishment, "than the purer manufactured drugs supplied to them through the Medical Department."

> In fact they insist that in some tropical diseases, notably elephantiasis, trained medical men have not been able to supply the medicine to effect cures, and their belief is strong in the value of their own drugs, the knowledge concerning which is a secret handed down from generation to generation among the descendants originally discovering the supposed palliative effect of the drug.[22]

I was pleased that the physician who would be assisting me was far more broad-minded than Dr. Hunt, the Rev. Turner, or his other medical predecessors in Samoa. My purpose in welcoming John, a physician Barbara and I had known for many years, was to help me to better understand Samoan healing systems. For a botanist without medical training, determining the anatomical analogue of, for example, the *toala* is difficult, but the question should be easily resolved by an M.D. When we learned that John would be attending a conference in Hawaii, we sent him an air ticket and an invitation to spend two weeks with us in Savaii. Because we had been friends for many years, Barbara and I eagerly anticipated his visit. But trouble started soon after John arrived at the airport.

"Well, John, you got your gamma globulin shot, didn't you?" I asked as we drove toward Apia. Most Samoans appear to have natural immunity to some forms of hepatitis, perhaps acquired during their childhood. But before the use of gamma globulin, a fourth of all *palagi* resident in villages contracted hepatitis.

"Paul, uh, I just ran out of time . . ." John began.

"I warned you that you would be exposed to hepatitis in Savaii," I said. "I watched Harold Moore, the world's expert on palms, die from hepatitis in Cornell Hospital after returning from fieldwork with me in Samoa. I spoke at his funeral—I don't want to speak at yours."

"I asked our surgical resident at the hospital," John replied. "He told me if I drank only bottled water and was careful what I ate, I would be O.K."

"In this culture, if you refuse to eat the food, you insult people. And there is no bottled water in the entire country."

I discovered that John's fear of injections had prevented him from obtaining the needed immunization, even though he routinely administered injections to hundreds of patients in his practice. However, after some coaxing, I convinced him to allow a nurse to give him an injection from my own private supply of gamma globulin, stored in a friend's refrigerator in Apia.

But John's fear of the needle was small compared with his fear of the people. Even though we hadn't yet boarded the ferry for Savaii, John acted as if he were behind enemy lines. He didn't want to try the local food. He didn't want to shake hands with the people. As we visited Epenesa Mauigoa, a Savaii healer who lived with her children near Apia, he was clearly uncomfortable.

Epenesa was a grandmotherly sort of person. She was short, about five feet two inches, and weighed perhaps 105 pounds. She had gray hair, dark brown eyes, and a lovely olive skin tone. She was quiet, both at home and in public, and did not, through her appearance or demeanor, arouse much attention. She was seventy-four years old, and her husband, an orator, was her elder by three years. They lived together in a small tin-roofed hut on the outskirts of Apia with her son Laiuni, a meteorologist at the Apia observatory, his wife, their nine children, and various members of their extended family who happened to visit. Her major avocations included daily attendance at worship services at the nearby temple, caring for her invalid husband, and knitting an afghan. But she was one of the most knowledgeable healers in all of Samoa.

"Epenesa," I said, "tell us about the *toala.*"

Epenesa confirmed much of what I had already heard in Falealupo: Movement of the *toala* can cause serious disease. As I related Leofo's reaction on seeing the body of the boy killed by dynamite, she concurred with the assessment of the villagers—he was lucky to be alive. Death from *toala* movement, however, is rare. More common is an internal ailment called *oso fa puni moa*, characterized by feelings of stomach blockage and pain. Her son Laiuni, who happened by, related how he had suffered as a youth from the ailment. "I told the *palagi* principal that I had to miss school because I had *oso fa puni moa,*" he recalled. "The principal wanted to know what that meant. I knew that *moa* means chicken in English, and puni means the *toala* is clogged, so I told him that my chicken was clogged."

"How did the principal respond?" I asked.

"He got mad," Laiuni laughed.

I asked Epenesa how she treated *toala* disease. She told me that she performed massage, and also used an herbal remedy manufactured from the bark of *Bischofia javanica,* the fruits of *Capsicum frutescens,* and the leaves of *Centella asiatica.* I typed the recipe in my laptop computer, and translated it for John. When I finished, Epenesa said there was an additional remedy she knew. "You can also use the *toi* tree [*Alphitonia zizyphoides*] just like the *o'a* [*Bischofia*] bark. The utility of *toi* bark is that it can cause the *toala* to return to its normal position and heal it. But the patient must not become chilled."

I was puzzled by Epenesa's use of the same word, *nonu,* for plants that were obviously different. She told me that the common type of *nonu* (*Morinda citrifolia*) is useful for treating skin rashes, while a plant she called *nonu fi'afi'a* (*Syzygium malaccense*), "the one with the insect galls on the leaves," is useful for treating internal ailments.

I told Epenesa that John was a doctor and asked if she could show him the location of the *toala.* Epenesa asked her husband to remove his shirt and lay on the mat. She pointed out the area below his navel.

"Is it the navel?" John asked while I translated.

"No—it's below there, deep inside. Sometimes when I put my hands on the abdomen, I can feel it like it has a pulse—I can palpitate it."

"Perhaps it's the inferior vena cava," John responded.

Epenesa asked what he had said. I told her that John thought it might be a very large blood vessel.

"You mean the one coming from the heart? No, it's not that," Epenesa responded.

"Is this a real thing or an imaginary thing?" John asked Epenesa.

"It's real," Epenesa replied.

"Her husband doesn't look very well," John noted. "He looks a bit jaundiced."

I translated for Epenesa. She responded that her husband was old and that his liver was slowly failing. She wanted to know if John could do anything for her husband. John listened to the man with his stethoscope, commenting that he had some obvious respiratory impairment. "In this setting, I really can't do anything," John said. "Tell her to get him to the hospital for a checkup."

John and Epenesa discussed anatomy for another half hour. John asked me if Epenesa owned a copy of *Gray's Anatomy* or had taken anatomy in school. She hadn't, but I assumed that the question indicated a positive appraisal of Epenesa's anatomical knowledge. As I was leaving, Epenesa asked me if I could get her some *Homalanthus nutans*, a small tree known as *mamala* in Savaii. She said it was a general remedy and also very important in treating *fiva samasama*, "yellowing fever." She then recited the recipe in typical healer shorthand:

E valu fasi la'au o le ateate ma se ogala'au o le mamala moni. O le mamala moni e lau lapotopoto o le mamala e pei o le maota lona tino ae laiti: auā o le maota e lau tetele. Palu lea i se ieie ae tunu ia puna se vai ona ligi lea i se ipu. E inu fa'alua i le aso. E sā mea suamalie ma mea lololo. Sā mea o le mea tau Popo.

Take eight stems of Wedelia biflora *and the trunk of a true* Homalanthus. *The true* Homalanthus *has round leaves and resembles* Dysoxylum *in its trunk but is far smaller and has smaller leaves. Macerate it into a cloth and plunge the cloth [like a tea bag] into boiling water. Decant the infusion into a cup. Sweets and greasy things are forbidden to the patient. Also forbidden is anything with coconut fat.*

I promised Epenesa that I would bring back some *mamala*.

On our way back to the Bethams' that evening, I drove John up into the mountains to visit the daughter of some rather poor

Samoans. Five-year-old Lolina, the daughter of one of the Bethams' plantation foremen, was born cross-eyed. I was concerned that without treatment she might go blind. Barbara and I had offered to pay for her and her mother to travel abroad to receive medical care. We hoped that John could tell us whether surgery might be advisable.

The road was rough, and my four-wheel drive took a pounding. Finally John spoke: "How can you do this?"

"Do what?"

"How can you bring me up here into this forsaken jungle and risk my life?"

I stopped the vehicle. "How am I risking your life?" I asked, genuinely puzzled.

"What if we get a flat? We'll be stuck here all night!"

I explained to John that there are no dangerous animals in Samoa and that the people are friendly—we were in no danger.

"Get me out of here!" John yelled. "These are filthy, dirty people and I want to go home now! Turn the jeep around!"

"John, let me get this straight. You have taken a physician's oath to help the sick. We are within two miles of a little five-year-old girl who may be going blind, and you want me to take you back before we even get there?"

"Paul, I'm not fooling with you. You may think it's fun hanging around with these witch doctors, but I think you're crazy. These people are savages—they don't have hygiene, they're full of parasites, and as far as I'm concerned they're relics of the Stone Age. Turn around, *now!* I want to get on the first plane out of here."

Without a word I turned the jeep around and drove John back to where we were staying. The next morning, we went to Apia to book his return flight. On our way out of the airline office, John saw some baskets displayed by a Samoan woman on the sidewalk in front of a store. "I need to take some souvenirs home. Ask her how much she wants for this basket."

I translated John's request. The woman smiled. "Large baskets like that sell for seven *tālā*."

"She wants seven *tālā*, John, about three U.S. dollars."

"Offer her two *tālā*," John ordered.

"This is not a barter culture. People ask what they believe to be a fair price, and then you should either accept it or reject it."

"Offer her two *tālā*," John repeated.

"I won't translate your request. I think seven *tala* is a very reasonable price for a large basket like this. It probably took her nearly three days to make it."

"If you refuse to help, then I'll just do it myself," John said.

John pulled out two *tālā* from his wallet and waved them in front of the woman's face. "Lookee here! Two *tālā!* Yes—I give you two *tālā* for this basket."

The woman shook her head. John grabbed another *tālā* from his wallet. "Lookee here! Three *tālā!* Last offer!"

The woman, now uncomfortable, again shook her head. Other Samoans on the sidewalk began to gather to watch the spectacle.

My patience snapped. "John, get back in the car! I'll get the basket for you."

I opened the car door and resisted the impulse to shove him in. After John was in the car, I went back to the woman.

"Please forgive my friend for his outburst. Neither of us wish to imply any disrespect to you or to your beautiful handiwork. My friend has not been feeling well and is returning to America today."

The woman smiled with generous, forgiving eyes.

"I like your basket very much and wish to buy it. Here are seven *tala* for the basket," I said, counting out the bills, "and an additional two *tala* as a gift for your patience."

"But, sir, the basket costs only seven *tala*. I couldn't take the extra money," the woman protested.

"Please accept it as a small gift to buy something for your children."

"Thank you very much, sir," the woman said.

I got back into the car. "How much did you get it for?" John asked.

"For you, John, it is free. My gift to you."

I drove John to the airport, put him on the plane, and headed for Savaii.

Fortunately, not all visitors share John's negative attitude toward the Samoans. After I returned to Falealupo, an American Peace Corps volunteer named Mark Muckerheide was hosted in the village by Fuiono Mase'ese'e. As time passed, Mark quickly picked up on Samoan customs, including bringing little gifts of food to Fuiono's family. A devout Catholic, Mark became a regular visitor to Fuiono Mase'ese'e's *fale* after mass.

One day, Lamositele asked me if I knew about Mark and Fuiono's daughter Amanda. I thought about the time that Mark had been spending with her family, his frequent food gifts, and suddenly realized that Mark had been conducting a formal Samoan courtship.

The next time Mark came to visit me, he spoke about Amanda. "I want to marry her."

"What about the Peace Corps—will they give you any trouble?"

"They don't care. A volunteer got married last year."

I smiled at this difference from my own experience being young and single in Samoa: Mormon missionaries are prohibited from any amorous involvement during their term of service. And, even had there been no prohibition, I was not romantically attracted to Samoan women. But then Westerners have long differed in their attitudes on this issue.

"Among a pretty large number of women I observed two or three of pleasing countenances," La Pérouse wrote in 1786. "Their shape was elegant; their arms well-turned, and nicely proportioned; and their eyes, countenances, and gestures, bespoke gentleness."[23] George Hamilton of HMS *Pandora* in 1791 was equally enthusiastic: "One woman amongst many others came on board. She was six feet high, of exquisite beauty, and exact symmetry."[24] Charles Wilkes, however, thought Samoan men far more handsome,[25] an opinion also expressed by John Erskine in 1849: "The inferiority of their beauty, compared to that of the men, is most striking."[26] A more evenhanded view was rendered by T. H. Hood in 1862: "The women are nice-looking, and have good strong figures, but cannot generally be called pretty."[27]

Through the years, there has been significant intermarriage between Samoans and foreigners, with the resultant "half-caste" community emerging as the major merchant class of Apia. And now, with the ease of air travel, many Samoans (usually men) who marry foreigners reside overseas. Mark had not yet determined his future plans, but I could not imagine a better match for him: Amanda Fuiono was lovely, intelligent, charming, and shared his commitment to service. Mark decided to marry her in the Falealupo Catholic church. Amanda seemed aglow at the announcement and began to weave a fine mat.

On my return from Apia, I was anxious to ask Pela about *Homalanthus*. She said that she used the bark of the tree to treat *tulitā*, an abdominal complaint. She directed me to the forest not far behind her *fale*, where *Homalanthus* sometimes grows up to five meters tall, but most often is much smaller. It grows along the edge of the rain forest and has beautiful spade-shaped leaves that have a slight whitish sheen on the bottom. One of the more striking features of the leaves is the long slender petioles that attach the leaf blade to the stem. Each leaf seems to hang suspended in the air by a tiny thread.

I was very interested in Epenesa's account because *fiva samasama*, "yellowing fever," sounded like an acute viral illness, perhaps yellow fever or hepatitis. I knew that there were very few pharmaceutical compounds known to be effective against viruses.

In the Falealupo forest I cut off some of the larger branches of a *Homalanthus* tree with my machete. Placing the chopped branches in a coconut leaf basket, I returned to our little *fale* to prepare voucher specimens. It is customary in ethnobotanical research to preserve a sample called a herbarium voucher specimen, so that if by some chance the plant is found to be active against a disease target, it can be re-collected. Proper preparation of a voucher specimen guards against misidentification and ensures that the botanical identity of the sample can be verified by other experts. The technology botanists use to preserve voucher specimens is much the same as that of Captain Cook's era: wooden plant presses and blotters. I took a branch and leaves of the plant, carefully flattened them between sheets of newsprint, placed the sheets between two felt blotters, and inserted the sandwich into my plant press.

The plant press is constructed of several large boards bound tightly with long adjustable straps. Between the different plant/ newspaper/blotter sandwiches are placed sheets of corrugated cardboard to allow air circulation and to ensure that the plants dry flat and straight. Since Samoa is so humid, drying and preparing samples is a rather laborious process—if not done properly, the samples quickly mold. The nearly constant 80 percent relative humidity in Samoa makes it necessary to dry the entire press over a kerosene stove. Barbara and I were always worried about this arrangement, since once in the field we had accidentally set a hut on fire while drying plants.

After setting aside some of the *Homalanthus* material for Epenesa, I prepared another sample for pharmacological testing by chopping up the bark and stem wood and packing them in an aluminum Sigg bottle. I then filled the bottle with 75 percent ethanol. Taking a solvent-proof marking pen, I wrote on the outside of the bottle: "#842 *Homalanthus acuminatus*, stemwood." I then made the following entry in the collection log:

Flora of Samoa
Island of Savaii

Homalanthus acuminatus
(Muell. Arg.) Pax
small tree 3 m. tall
native name: mamala.
native use: medicine–bark
used for incontinency.
leaves and bark used
for yellow fever.
Paul Alan Cox #842

Although *Homalanthus* was not used by the healers to treat cancer, I placed the aluminum bottle in a box with the other samples I had collected for screening by the National Cancer Institute in Bethesda, Maryland. The fact that the healers believed a plant to show any healing property was sufficient for me to want to screen it on the off chance that it might have some undiscovered anticancer properties.

Preserving plant samples in fluid is unusual; most plants collected for pharmacological testing are simply dried in the sun. Since Samoan healers use only fresh material, I worried that some healing properties might be lost on drying. But I had agonized over the choice of solvents for a variety of reasons, including safety: Samoans are relatively unaware of the dangers of modern chemicals. Given the toxic nature of most organic solvents, the results of naive experimentation would almost certainly be lethal. So I chose to use ethanol. At high doses it is extremely toxic, but at low doses it might be survived by a curious child with no more serious consequence than a severe hangover. But this choice brought me into conflict with my own professional code of ethics.

Near our *fale* Lamositele had built a shed with walls that had at one time served as a small store. As soon as I returned from Apia, I placed the jerry can of ethanol there. I was carefully pouring ethanol from the large jerry can into the small aluminum sample bottles when Lamositele approached me. "What are you pouring into the bottles?" he asked.

"It is a toxic chemical that I use to preserve my plant samples with," I truthfully replied.

"Funny," he said. "It smells like alcohol to me."

I thought quickly. Samoans, in my opinion, are saints, but they are also human, particularly where alcohol is concerned. If word spread through the neighboring villages that I had in my possession a ten-gallon can of ethanol, someone might attempt to steal it. Some of the village youths might attempt to experiment with it when I was gone. Such experimentation could lead to disaster, since few people in Savaii had ever consumed anything with a greater alcohol content than beer. Even a liter of pure ethanol might prove lethal if consumed in a single sitting. Even worse, the village chiefs' council might ask me for some of the alcohol, and since I was duty-bound to honor their decisions, I would feel compelled to comply. I thought back to the time as a missionary when I had watched a drunken Samoan demolish his house and attempt to kill his wife and child. How could I introduce such turmoil into Falealupo? Possessing such an intoxicant gave me a grave responsibility.

"Lamositele, this substance is an extremely toxic chemical used to preserve plants."

Lamositele turned away from me without a word. After he left, I felt a dull anguish. By lying to Lamositele, I differed little from the colonialists who saw indigenous people as savages, or children, or otherwise incompetent to make their own decisions. As a nineteen-year-old, I had come to Samoa to serve and learn from a people I respected, but now I had returned as a Ph.D. little different from the Reverend George Turner, determined to save Samoans from their "heathenism." Was I any better than Dr. John, whom I had returned to the airport so recently? At least he could claim ignorance. I promised myself that never again would I betray the trust of a Samoan.

10 cm.

Artocarpus altilis

CHAPTER 4

Return

*It was an extremely
charming sphere, the abode
of all the virtues.*

Joseph Conrad,
"The Return"

Nighttime in Falealupo. I listened to the hush of a gentle rain on the thatched roof and contemplated the changing mosaic of sounds that evidence the shifting pattern of village dominion in Samoa. At dawn a village belongs to the infants, whose persistent cries are soon quieted by their mothers' milk and tender attentions. The village of the morning belongs to the chiefs, whose eloquent kava speeches fill the *malae,* or village common, with majesty and beauty. The village of the afternoon is contested by the voices of women weaving mats, the laughter of children on the beach, and the excitement of a returning fishing party. At dusk, the village belongs to God as each family, regardless of religious preference, gathers for hymns and prayer. As evening falls, the gentle sounds of slack-key guitar and quiet singing announce the ascendancy of the village youths, who huddle in small groups beneath the breadfruit trees. As their voices slowly subside, the chirping of geckos, lizard acrobats that cling with suction-cup feet to posts and ceilings, fills the air. Later, the night is punctuated by the sounds of domestic dogs, who delineate their territories through both growled threat and infrequent violence. The kapok and guava trees are then quickly claimed by troops of quarrelsome flying foxes noisily jousting for roosting places. Finally, just before the dawn, the vast feudal kingdoms of the mighty roosters are noisily proclaimed even as first light steals their nocturnal splendor.

The next day was Sunday. Instead of driving Filiva'a, the Congregational minister, to his services along the beach road, I decided we should investigate the inland portions of Tufutafoe village, since I had never been there before. We headed inland to the main paved road and thence to Sataua to pick up the Mormon branch president and his family. We then drove back to Neiafu and turned down a dirt track leading to Tufutafoe. As soon as we started downhill on the dirt road, I was startled to see the litter of stumps and downed logs and rough, jagged scars in the red soil, stretching for miles in a vast swath toward the sea. Large cracks in the soil bore silent witness to the forces of erosion that had been unleashed on the once undisturbed forest.

Both ministers expressed surprise at the devastation. We drove two miles through the scarred and desolate landscape until we reached the village by the sea. Since we arrived early, Filiva'a invited us into the *fale* where the village leaders were gathered.

After responding to the formal entrance rhetoric and being introduced to the chiefs, I asked the question that had been on my mind ever since we had turned off the main road.

"When did the loggers come?"

"About a year and a half ago," one of the chiefs answered. "They cut about twenty thousand acres."

"Did the village receive payment?"

"Forty thousand *tālā*."

"Does the village consider that fair?"

"We did at the time."

"That's a lot of money," I said. "But how much does it amount to per acre?"

"Two *tālā*."

"How many big trees do you think there were in an acre of the rain forest?" I asked the chiefs.

"Around a thousand or so," one of the chiefs responded.

"In that case, the loggers paid the village two *sene*—one U.S. cent for ten trees. Tropical hardwoods fetch five hundred dollars per cubic meter on the world market," I said. "After milling, a sizeable tree can be converted into ten to twenty cubic meters."

The minister said nothing, but looked glum. A chief explained that the village didn't actually sell the forest to the loggers, but merely leased it to them for a twenty-year period. "But no one would want it, the way it is now," he lamented. "We didn't know they would scrape the soil off. Our springs have dried up, and we have a hard time even finding drinking water. None of us knew it would be this bad."

The other chiefs nodded in assent.

I returned by the coastal track to Falealupo. While I drove, I considered the logging of the Tufutafoe forest. The village had freely accepted the company's offer, but they acted in ignorance of the true value of their timber. Even if sold locally rather than on the world market, their logs would have fetched far more than the two *tālā* per acre the village received. And the village had acted without understanding the negative impact the logging would have on their village environment and their lives. Although the logging that occurred in Tufutafoe was perfectly legal, I questioned whether the village leaders truly had exercised informed consent.

After I concluded ferrying the ministers about, I returned to our *fale* for *to'ona'i,* or lunch. Barbara enjoyed the bananas baked in the *umu,* particularly the ones that were slightly burnt. All of the family enjoyed the baked breadfruit, but it took some coaxing to get our older children to try the octopus baked in coconut cream. Paul Matthew decided that it tasted like lobster. Hillary, much to the delight of the Samoans, was indiscriminate in her eating habits. She seemed to enjoy everything they put before her.

That night, which was moonless, I retrieved a cyanamide light stick from my backpack. By attaching a light stick, which glows pale green for eight hours, to the end of a staff, it is possible to navigate trails in complete darkness. But that night our children and the neighboring Samoans used the stick to play catch on the white sand in front of our *fale,* with the stars shining brightly overhead.

In some ways, Barbara and the children were becoming integrated into Samoan society more easily than I was. Each morning, our children picked leaves off the sand, swept the *fale,* and engaged in other chores with Fa'asaina's children. After completing their schoolwork, the children swam in the tidal pool with the village children. They were taught how to roast cocoa beans, how to weave baskets from coconut leaves, and how to sing Samoan songs. Barbara and Fa'asaina became very close, often going together to the swamp to wash clothes and visit with the other village women. But my presence had an impact on the flow of things. In my absence, Barbara and the children were able to eat informally in the cook hut with the rest of Pela's family, but when I was present, meals were served on banana leaves in our *fale.* When I was gone, there was a stream of village children through our house, but when I was around, only

Pela's grandchildren were allowed to visit. Just as some expatriates in Tokyo report that their fluency in the language causes wariness among the Japanese, it appeared that my Samoan language ability distanced me in some way from the normal flow of village life.

The next week, I continued my explorations of the Falealupo forest by hiking three miles up the road from our *fale* to a large volcanic cinder cone called Mauga Fuionō, "Fuiono's mountain." If there were Samoan flying foxes in Falealupo, they would roost on the mountain. I had studied flying fox pollination for my doctoral dissertation, and was anxious to see how many of the animals remained in Falealupo. Before leaving the road, I decided to ask permission of the family who lived nearby. A small path through cassava bushes and variegated-leaved *Jatropha* brought me to a small *fale*. It seemed almost Japanese in aspect—flat volcanic rocks surrounding the *fale* formed a kind of patio. Between the rocks flourished the type of small-bladed grass used to seed golf greens. A handsome middle-aged woman tending two small children saw me and spread out the mats of welcome. Her name was Uefa, she said, and she had recently been widowed. I asked if I might walk through what appeared to be her small family plantation to explore the cinder cone. Uefa agreed, but cautioned me that the path was rough. Had she ever seen any flying foxes in the forest?

"Yes," she replied. "A pair lives on top of the mountain. I see them flying in the afternoons."

The climb was hard and the day was hot. After ascending a few hundred meters through a small plantation of bananas and *ta'amu,* an edible plant similar to taro with giant "elephant ear" leaves, the path became more imaginary than real. I pulled the machete from my knapsack and used it to chop the vines and small saplings that obstructed the trail. I quickly climbed through secondary forest between *Macaranga harveyana* trees (known in Samoan as *lau papata*), whose large leaves, long petioles, and flimsy trunks resemble those of Central American *Cecropia* and other species adapted to colonize gaps in the rain forest. Climbing still, I entered the primary rain forest, characterized by large fig trees, towering *Dysoxylum maota* trees, and a genus of Sapindaceae, *Palaquim,* whose fruit is a favorite of flying foxes. Reaching the summit, I sat on a ledge overlooking the Falealupo peninsula and studied the configuration of the land below.

The agricultural activities of the Falealupo villagers seemed to be limited to a small corridor surrounding the village and the road leading to it. From the mountain, the rain forest continued west for two miles to Fagalele, a secluded beach used as refuge by fishermen in a storm. To the east, the high interior peaks of Savaii vanished into the clouds.

Suddenly there was a motion in the sky off to the south—a flying fox soaring on the afternoon thermals. I recognized it from my doctoral studies in Samoa: *Pteropus samoensis,* with black wings spanning nearly four feet, resembling a giant hawk above the forest. Only a few hundred meters to its side was another flying fox, likely its mate. They slowly turned, made some broad swooping flaps to gain altitude, and then flew straight toward me.

When I returned to our *fale* that evening I told Barbara what I had seen.

"Were they the diurnal species or the nocturnal species?" she asked.

"They're definitely *Pteropus samoensis,* the diurnal species—with no neck marking, but one of them had a white patch on its forehead."

The children were excited. As a family we were very concerned about the fate of flying foxes. We knew their numbers were declining rapidly, perhaps approaching extinction.

That night in the *fale* our children played Ludo, a portable board game, with some of Fa'asaina's children. Sixteen-year-old Lui seemed particularly enamored of the game, but grew frustrated when he learned that he couldn't arbitrarily place the pieces to his own advantage. After dinner, I found Lamositele preparing his gear for a midnight spearfishing excursion. Both his father, Lilo, and his brother, Lilo Manuele, were subsistence farmers. But Lamositele had decided as a boy that he would pursue a different path. "It was hard in the beginning," Lamositele said. "My father didn't fish and there was no one to teach me. So I started watching the other fishermen very closely, helping them whenever possible. Day by day, I learned a bit more: how to dive deep, how to dive at night, how to shoot my spear and not lose it, how to trawl from a canoe, and many other things. But the most important thing I learned was that to be a fisherman you must learn to master your fear."

"Why is it frightening to be a fisherman?"

"I was spearfishing outside the reef with a friend one day. He motioned for me to swim to him. I looked down and saw several very large stingrays, perhaps six or seven of them, stacked on top of each other, lying motionless on the bottom. Perhaps they were mating. My friend dove down, close enough to touch them. He held his breath and lay there above them, watching."

Lamositele pushed his cocoa mug forward on the rough mat, indicating that he was finished with it.

"I was staring at the stingrays below, when all of a sudden, I saw another one approach my friend from behind. He couldn't see it, and it came so rapidly toward him, he didn't know it was there— until it dropped down on him, clasping him with its wings."

Lamositele gestured with his hands, showing how the ray had embraced his friend.

"So there he was, sandwiched in this pile of stingrays. I was terrified—what could I do? I grabbed my spear and swam down. I knew I had to shoot the ray on top, but I was afraid of its stinger."

"*Ua sola le fai ae tu'u le fotu,*" I recited. "The stingray has left, but its stinger remains."

"That's very good," Lamositele said. "Do you know what that proverb means?"

"I think it refers to people who have done bad things, such as spreading gossip. The evil continues long after the perpetrator has fled."

"That's correct," he said. "When you're attacked by a stingray, the tip of the stinger breaks off inside you. It continues to move through your body until it eventually reaches your heart. Long after the ray has left, its stinger can still kill you."

"So how did you save your friend? Surely he couldn't have held his breath for very long."

"He was in trouble," replied Lamositele. "The problem was in finding the exact spot to spear. I had to get close enough to hit the area between the head and spine, but at that close range, if I missed, the stinger would whip up and get me."

"So what happened?"

"I dove to a foot above the ray, aimed at the neck point, and shot."

"And then?" I prompted.

"God guided my spear."

Near the road I had taken the previous day, I had seen a steel pipe that had once provided water to Falealupo. A later bacteriological check of the Falealupo water supply showed it unfit for consumption, so when the pump finally broke, the government did not repair it. There was, of course, plentiful brackish water in the swamp for washing, but during the dry season the villagers, including ourselves, had to depend for drinking water on the tank that drained the roof of the Catholic church.

A few months before our arrival, the government had launched a program in Asau subsidizing the construction of concrete water tanks, but the cost—$700 U.S.—was still beyond the means of most Samoan families. So I drove the six miles to Asau, and with the noise of the sawmill in the background, paid to have two large pre-fabricated water tanks installed in Falealupo. Perhaps I could later be reimbursed from my National Science Foundation award. I asked that the tanks be delivered to Pela and Lilo's *palagi*-style house, together with the rain gutters necessary to drain their roof into the tanks. The arrival of the truck with the tanks was a big event in the village. Many people gathered to see the thousand-gallon tanks being placed upon the rock foundation that Lamositele and I had built. Within an hour the white plastic rain gutters had been attached. A fortuitous light shower that night permitted us to share a wonderful moment in the morning—turning on the faucet at the bottom of the tank and seeing pure rainwater emerge.

The tanks permitted all of us not only to enjoy clean drinking water, but also to shower at our convenience, regardless of the tides that influenced the salinity of the village bathing pool. An added benefit was hot water. We had arrived from the United States with some of the black vinyl bags sold to backpackers for solar-heated showers. In the morning, I filled the shower bags from the tank, and placed them in the sun to heat up.

It gave me particular joy to supply water for Pela's husband's bath. Bathing in hot water was his favorite memory of New Zealand, the only overseas destination he had ever visited.

"It was wonderful to see my grandchildren and to be with my family there," Lilo claimed, "but the bathtub was about the only place I could get warm. My daughter had a bathtub long enough for me to stretch out in, so I used to lie there, and let the hot water come up just below my nose," he said, gesturing with his hand to show the height of the water. "And then, you know what I'd do? I'd fall asleep!" he said with a sheepish grin.

Fa'asaina, by contrast, was surprisingly resistant to the idea of using hot or cold tank water, at least for washing. She claimed to prefer to wash clothes in the swamp. Why would Fa'asaina want to carry her clothes to the swamp and sit in the muddy shallow water when she could have hot fresh water right at home?

"I've gone with her regularly to the swamp," Barbara volunteered when I mentioned my puzzlement. "I have to admit it's rather fun. It's pleasant to sit in the cool water in the heat of the day. All of the village women are down there. There's a lot of visiting, just like quilting bees back home in the old days. It's a kind of social club for women. Fa'asaina works hard for her family all day, and it's her sole opportunity to relax. Haven't you realized that it doesn't actually take two hours every day just to do the wash?"

Barbara's explanation of the swamp as a social club solved the conundrum, but Fa'asaina's refusal also served as an example of the long-standing Samoan ambivalence toward Western technology. Early European visitors were nonplussed by the Samoan rejection of beads, iron implements, and other trade goods. "Though the canoes of these islanders were skillfully executed, and afforded proof of their ability in working in wood, we could not prevail on them to take our hatchets, or any iron tools," lamented French explorer La Pérouse in 1787.[28]

Surely a people who produce such beautiful handiwork should admire Western craftsmanship. Dutchman Jacob Roggeveen, who in 1722 became the first European to sight Samoa, reported that Samoan seagoing vessels were "not hollowed-out trees, but made of planks and inner timbers and very neatly joined together, so that we supposed that they must have tools of iron."[29]

But Fa'asaina's attitude again evidenced that Samoans prize decorum over convenience and relationships over possessions. The importance of both in the culture was reinforced the next morning when I visited Seumanutafa Siaosi, who is *fōfōgau,* a or bone setter, one of the few Samoan healing specialties practiced by men. As Seumanutafa, perhaps sixty years old, gently rocked a baby on his lap, we discussed Samoan healing. Suddenly, a tourist stepped into his *fale.* She walked across the mats with her shoes on and, without a word, began to snap flash photographs of Seumanutafa. Not only was her behavior stunning, but her very presence in Falealupo, one of the most remote villages in Samoa, was amazing. Despite her violation of nearly all Samoan customs, including her immodest (by Samoan standards) beach shorts, her failure to remove her shoes upon entering, her standing in the presence of a high chief, and her taking of photographs without permission, Seumanutafa remained unperturbed.

"Don't you mind this tourist taking your photograph?" I asked.

"Is such behavior acceptable in *palagi* culture?" he responded.

"No. Let me speak to her."

I turned and addressed the woman in English. "I don't know if you think it is proper to enter someone's home without an invitation, but don't you think you should ask permission before you take someone's photograph? This man is a high chief of this village."

"Well, ask his permission then," she said with a German accent.

"She seeks permission to photograph you," I relayed.

"Tell her that she can continue," Seumanutafa replied. Then, with a twinkle in his eye, he added, "Wait—why don't you say 'no' and let's see how she reacts."

I turned to translate. "High Chief Seumanutafa declines your request to be photographed."

With a scowl, the woman turned and stomped off toward the road where her vehicle waited. As soon as she was out of earshot Seumanutafa and I burst into laughter.

"I've got an idea," I said. "After dinner why don't we drive to the Vaisala Hotel in Asau. That must be where she is staying. Precisely at midnight, we will burst into her room and take flash photographs of her in bed."

At this suggestion, the smile drained from Seumanutafa's face. "No, we could never do that," he said quietly. "*E le tagata paopao tatou:* We are not savages."

Polynesians in general, like Seumanutafa in particular, pride themselves on this assertion of shared humanity. When questioned by European explorers, the islanders used cognates of a single term to identify themselves: *tagata, kanaka, kanak:* "people." Colonial and even more recent visitors have arrived with quite different opinions, however—views that still live on among people like my doctor friend and the German tourist. The second white visitor to Samoa, La Pérouse, found the Samoans in 1787 to be "a barbarous people":

> To others I willingly leave the task of writing the uninteresting history of these barbarous people. A stay of twenty-four hours, and the relation of our misfortunes, are sufficient to make known their ferocious manners.[30]

La Pérouse proclaimed the Samoans "ferocious" because of an altercation between his crew and the residents of Asu village in Tutuila, which resulted in the death of eleven French sailors. He claimed that the Samoans attacked the sailors without provocation, but other witnesses reported that a sailor shot a Samoan dead when he attempted to steal a nail. The Samoans reacted by throwing stones, while the French indiscriminately discharged their muskets into the crowd of men, women, and children. The number of Samoans who perished was never reported.

Because of La Pérouse's warning, European vessels avoided Samoa for over forty years. This was probably a blessing for Samoa, since, unlike Hawaii and Tahiti, the archipelago was colonized and politically subjugated by *palagi* only late in the eighteenth century.

Missionary John Williams was clearly either unaware of the La Pérouse incident or not dissuaded by his warning. In 1830 Williams found the Samoans to be anything but ferocious:

> An immense crowd had assembled to witness, I believe, the very first Englishmen who set foot upon their shores. . . . As

we were walking along, having intimated to the young chief
that I was exceedingly fatigued from labouring the whole of
the day in the boat, he uttered something to his people, and
in an instant a number of stout fellows seized me, some by
my legs, and others by my arms, one placing his hand under
my body, another, unable to obtain so large a space, poking
a finger against me, and thus, sprawling at full length upon
their extended arms and hands, I was carried a distance of
half a mile, and deposited safely and carefully in the presence
of the chief and his principal wife.[31]

With the extraordinarily warm reception accorded to Christian
missionaries by Samoans in the 1830s, the informal boycott of
Samoa by sailors ended. A decade later, Captain Wilkes of the
U.S. Exploring Expedition found the Samoans to be "kind, good-
humored, intelligent, fond of amusements, desirous of pleasing
and very hospitable."[32] In his appreciation of the Samoan character, Wilkes was not alone. "Indeed, from the first, their politeness
and good manners struck us as equal to that of any country we had
ever seen," John Erskine wrote in 1849.

From what I have already seen and heard, I do not believe
there is a country in the world, however strictly the laws
may be administered . . . where a white man can live with
such a sense of safety, and certainty of a supply of food, as
in these islands.[33]

Some later visitors, however, particularly those frustrated by
the intricacies of Samoan politics, published far more negative reviews of the Samoan character. Colonel Stephen Allen, a colonial
administrator in the 1930s from New Zealand, claimed that "the
Samoan never grows up, but always retains the mind and intellect
of a child, reasons like a child, and behaves like a spoilt child—as
he actually is."

The main thing to remember is that the Samoan has never had
to think for himself. . . . The Samoan is consequently destitute
of reasoning power and incapable of connecting cause and
effect. . . . In fact the Samoan in his early twenties may be
considered the mental equivalent of a European boy of twelve
or fourteen and he never advances beyond that stage.[34]

Allen's comparison of Samoans to children was of course not
novel. The view of Polynesians as uninhibited children, devoid of
Western strictures on sexual morality, has been a theme in many
accounts since Captain Cook's day. The thesis reached its apogee in

the most famous book ever written on Samoa: *Coming of Age in Samoa: A Psychological Study of Primitive Youth for Western Civilization,* published in 1928 by Margaret Mead. Her view of Samoans as primitives was made clear at the outset:

> For such studies the anthropologist chooses quite simple peoples, primitive peoples, whose society has never attained the complexity of our own. . . . In complicated civilizations like those of Europe, or of the higher civilizations of the East, years of study are necessary before the student can begin to understand the forces at work within them. . . . A primitive people without a written language present a much less elaborate problem, and a trained student can master the fundamental structure of a primitive society in a few months. . . . So, in order to investigate the particular problem, I chose not to go to Germany or to Russia, but to Samoa.[35]

During an ethnobotanical survey in 1978, I stayed in the villages of Ta'ū island, an island in the Manu'a group made famous as the site where Margaret Mead performed her fieldwork beginning in November 1925. Though 180 miles east of Savaii, Ta'u is peopled by Samoans of nearly identical language and culture. Having read press accounts of Mead's popularity in Samoa, I was surprised to find the villagers reluctant to discuss her. Some were old enough to have been village youths during the seven months of her fieldwork. Those who would discuss her told me something remarkable: Margaret Mead did not speak Samoan. This was a surprising assertion, since she claimed competency in the language:

> Speaking their language, eating their food, sitting barefoot and cross-legged upon the pebbly floor, I did the best to minimize the differences between us. . . .

> With a few unimportant exceptions this material was obtained in the Samoan language and not through the medium of interpreters. All of the work with individuals was done in the native language, as there were no young people on the island who spoke English.[36]

While she was respected in scientific circles as the intellectual heir to anthropologist Franz Boas, Margaret Mead's place in popular culture was assured not by her professed linguistic ability, but by her claim that sexual promiscuity eased the trauma of adolescence in Samoa. "Was the knowledge of sex and the freedom to experiment a sufficient guarantee to all Samoan girls of a perfect adjustment?" Mead asked in her book.

In almost all cases, yes. . . . The opportunity to experiment freely, the complete familiarity with sex . . . make her sex experiences less charged with possibilities of conflict than they are in a more rigid and self-conscious civilization.[37]

The "rigid and self-conscious civilization" referred to, of course, was America, where "our life-histories are filled with the later difficulties which can be traced back to some early, highly charged experience with sex." Mead explained that things are quite different in Samoa:

The Samoan child faces no such dilemma. Sex is a natural, pleasurable thing; the freedom with which it may be indulged in is limited by just one consideration, social status. . . . All this means that casual sex relations carry no onus of strong attachment, that the marriage of convenience dictated by economic and social considerations is easily borne and casually broken without strong emotion.[38]

Casual sex in the islands! No onus of attachment! Marriages casually broken without strong emotion! Margaret Mead had stumbled on a sure-fire formula to raise any book on Polynesia to instant best-seller status. But during my fieldwork I found no evidence that the Ta'ūans are libertines. Of course, since I arrived half a century later to study ethnobotany rather than adolescent development, my failure to discover the sexually liberated society described by Margaret Mead cannot be considered a refutation of her work. But her claims, made as they were by a twenty-three-year-old graduate student during her very first field study, perhaps should have been considered more skeptically, particularly since they contradicted so many previous accounts of Samoa.

"Here there is no indiscriminate intercourse, the marriage tie is respected, and parents are extremely fond of their offspring,"[39] Captain Wilkes recorded in 1845. He later observed that "venereal disease does not exist at Tutuila, and is hardly known in the other islands. This serves to prove how great a superiority this island possesses over Tahiti in the chastity of its females, who in general observe their marriage vows with strict fidelity."[40] Wilkes noted the penalties Samoans placed on marital infidelity and sexual promiscuity: "Adultery was formerly punished by death, and is very seldom committed. Among single women, intercourse with a Samoan before marriage, is a reproach, but not with transient foreigners."[41]

Even more puzzling was how Margaret Mead could have failed to recognize the importance of religion in Samoa, relegating it to a single paragraph in her book. The large and elaborate London

Missionary Society churches of T'aū, some over a century old, are the most prominent structures on the entire island. Mead failed to note that the Ta'ūans were far more dutiful in their observance of Christianity than the "rigid and self-conscious civilization" she left behind during the Roaring Twenties. But most perplexing to me was the lack of affection on the part of the otherwise friendly villagers for a person who brought their small village international fame.

"So what's the deal with Margaret Mead?" I asked a young Samoan who held a master's degree in sociology. "None of the chiefs or older women seem willing to say much about her."

"Oh, her," he said. "The old people are really angry at Mead."

"Why? I thought they liked her."

"That was before they knew what she wrote about them," he said.

"Mead argued that Samoans have an easier time passing through adolescence than *palagi* teenagers do," I offered.

"Maybe they do, maybe they don't," he said. "But it blew the old people's minds when one of the village kids came back from college several years ago and translated a few excerpts from her book for them."

"You mean that for all these years the villagers didn't know what she had written about them?" I asked.

"No. They all assumed that it was very complimentary. But when they heard that she claimed they were having sex with nearly everybody in sight, they burned her book."

"Why do you think Mead wrote those things? You don't think she just made them up, do you?" I asked.

"Samoans joke about anything, particularly with foreigners. Somebody just fed her a bunch of baloney and she believed it."

A few months after my visit to Ta'u, Margaret Mead died, properly eulogized as one of the great figures in twentieth-century anthropology. But in 1983, five years after her death, her name again made headlines. Newspapers were filled with wire stories about the "debunking" of Margaret Mead by Australian anthropologist Derek Freeman.

In *Margaret Mead and Samoa,* Freeman argued that Mead did not understand the Samoan language, was misled by her interpreter, and as a result largely misinterpreted Samoan society. Placing Mead's work in the context of the "nature versus nurture" debate of the 1920s, Freeman argued that Mead had created the myth of a Samoa characterized by sexual promiscuity, lack of jealousy, and open marriage to serve her own views. To buttress his attack, Freeman relied not only on his own experience in Sa'anapu village in Upolu (where he took a chief's title), but also on the fieldwork of anthropologist Eleanor Gerber in American Samoa, who found that "premarital chastity" was the rule. In the days of their grandmothers, her informants told her, observance was far stricter, with "all daughters virginal."[42]

Freeman also took Mead to task for asserting that Samoa was a primitive and simple society. "As any one who cares to consult Augustin Kramer's *Die Samoa-Inseln,* Robert Louis Stevenson's *A Footnote to History,* or J. W. Davidson's *Samoa mo Samoa,* Samoan society and culture are by no means simple and uncomplex," Freeman writes. "They are marked by particularities, intricacies, and subtleties quite as daunting as those which face students of Europe and Asia."[43]

Yet I was troubled not only by the timing of Freeman's posthumous assault on Mead and the tremendous prepublication hype that Harvard University Press released to push book sales, but also by Freeman's own view of Samoans. His vision of the Samoan personality seemed scarcely less simplistic than Mead's. "Samoans are authoritarian, competitive and touchy," Freeman said in a speech reported by the *Honolulu Star Bulletin.*

> They are brutal in war. They have higher rates of assault
> and rape than most of the world. Their adolescence is
> tumultuous and of all the cultures that anthropology is
> conversant with, it is the one with the greatest
> emphasis on the cult of virginity.[44]

Freeman was remarkably defensive of his own analysis and "accused anthropologist Brad Shore, who mildly criticized [him], of making 'a travesty of the truth.' He added that Shore's book on Samoa confuses two key words and 'ought to be issued with an *errata* slip in the front.' He later called Shore's book 'the biggest blue [mistake] in the history of anthropology.'"[45]

Noncompetitive, lenient, and somewhat promiscuous. Authoritarian, competitive, and touchy. Margaret Mead's and Derek Freeman's descriptions of Samoans seem worlds apart. The longer we lived in Falealupo, the more I wondered if what one sees in Samoa is merely a reflection of one's own preconceptions.

My own gaze into the Samoan mirror, however, was becoming clouded. Although I appreciated the kindness of Pela's family and the villagers to our family, I was becoming increasingly uncomfortable with the deference accorded to me. My food was served on a separate tray. Villagers began to greet me with words reserved for chiefs. Shares of pigs slaughtered for village functions started arriving at our *fale* unsolicited. I was now occasionally asked to speak for the village on ceremonial occasions. In short, the hospitality extended to me began to exceed that ordinarily offered to foreigners, even though I had done nothing to merit such consideration. It had been my hope that within a few weeks of our arrival, the novelty of a *palagi* family taking up residence in the village would have worn off. I wanted to reduce distance between the villagers and myself, but I feared that I was coming to be seen as a character out of a Joseph Conrad novel: "He talked to me (the second white man he had ever seen) with confidence, and most of his talk was about the first white man he had ever seen. He called him Tuan Jim, and the tone of his references was made remarkable by a strange mixture of familiarity and awe."[46] I had returned to Samoa as an ethnobotanist. I did not want to become Lord Jim.

As a young missionary, I had fit into a well-established niche within Samoan society—*faifeau,* or minister of religion—and my relationship to the villagers was clear at all times. But my return to village life in Samoa this time brought with it ambiguity in my status. In some respects, it would have been easier for both me and the villagers had I not spoken any Samoan at all—at least then I could have been considered a tourist. But my ability in chiefly language, taught to me so long ago by Aumalosi, had unintended consequences in a culture that places such high value on oratorical skill: rather than being accepted as ordinary, the longer we stayed, the more my status in the village increased.

The issue came to a head one night after dinner. Lamositele, his brother Lilo Manuele, and Silia Tusi, a village orator, came in and sat by me in the *fale*. Lamositele poured hot Samoan cocoa for each of us into tin cups.

"You speak Samoan very well," Silia began.

"Thank you."

"You speak better than some of our chiefs," he continued. His assessment was overly generous, but he stopped my protest. "We think you should take a chief's title."

I glanced at Lamositele and his brother, Lilo Manuele. They weren't joking. I tried to think of a way to politely refuse their offer. Not only would a title make my ethnobotanical work much more difficult by increasing, rather than reducing, my distance from the villagers, but I also believed that Lilo Manuele and Lamositele overestimated my rhetorical ability. But perhaps the deepest reason for avoiding a title was the one that I was least willing to articulate, particularly to the Samoans who had given me so much: I just didn't want that level of responsibility for the village and for the family.

Nearly every Samoan extended family has a series of chiefly titles in their genealogy that can be conferred on family members. Only one family member can bear a given title, but the same title can sometimes occur in other families. Thus, in Falealupo, men from three different families hold the Fuiono title. Each title is associated with a particular parcel of land over which the chief becomes steward, acting on behalf of the extended family in allocation of resources. When the title holder dies or becomes infirm, the extended family is convened to determine who should receive the title. Selection is a time-consuming process, since every member of the extended family has a say in the matter. The Samoans say, *"O le ala i le pule o le tautua"* ("The way to leadership is service"). Only individuals who have shown a willingness to serve the family and village are chosen.

On assuming a title, a new chief becomes part of the village council, and is expected to work in concert with the other chiefs for the betterment of the village. Titles are of two types: *ali'i,* or high chief titles, and *tulafale,* or orator titles. In village councils, high chiefs function like court justices, adjudicating the merits of issues that arise, while orators (sometimes called talking chiefs) function like lawyers, arguing cases, manipulating the politics of an issue, and extracting payment in the form of fine mats, pigs, canned food, and cash from those less adroit.

I turned to Lamositele and Lilo and asked them again why they wanted me to become a chief.

"You speak like one. You're more than just a temporary visitor, and the village needs to establish some way to relate to you in an ongoing fashion."

"I don't feel qualified to become a chief. *O le ala i le matai o le tautua:* The way to chiefdom is service," I recited. "I haven't really done anything to serve the family or the village. If the village wants to relate to me in an ongoing way, then make me a *taule'ale'a,* an untitled man."

Lamositele, Lilo, and Silia looked perplexed.

"We don't make untitled men. They are just born," Lilo explained.

The three chiefs left me that evening, but the subject they had raised did not die. The next week, while I was pressing plants, Lamositele approached me.

"Koki, I think I understand why you don't want a chiefly title, but I disagree with your reasoning. Everybody knows that you speak better respect language than many chiefs, and the few things you lack concerning kava speeches can be easily taught. And you have served the village. Every Sunday you drive the Catholic minister to mass, you take the Congregational minister over to Tufutafoe to hold his service, yet you yourself are a Mormon. You've taken villagers to the hospital late at night, you've built these concrete water tanks for us, and you have done many other things for the village. And there is one thing that you don't seem to understand: your contributions make it increasingly awkward for us to deal with you as an outsider. Other _palagi_ people have accepted chief's titles. Everyone would be more comfortable if you would truly become a member of the village by accepting a chief's title."

"But those _palagi_ were different," I countered. "Visiting dignitaries like the Prime Minister of New Zealand accept a title as a pleasant honorific. I take the position too seriously to consider it just a symbolic honor, and I understand it well enough to know that I am unqualified to be a chief. If you need to relate to me in some formal way, then make me an untitled man."

"You don't understand. We don't confer the status of _taule'ale'a_ on people—you are born to it," Lamositele said.

"Lamositele, it is your culture. How you go about making an untitled man is your business."

After considerable discussion, the village and Pela's family devised a little ceremony to make me an untitled man. We had a nice dinner together, at which they presented me with a roasted pig, and afterward a beautiful and specially woven fine mat.

"And with this," Pela's son Lilo Manuele said, "I now pronounce you officially an untitled man of Falealupo village. You are now truly the flesh and blood of Falealupo, just as if you had been born here."

I accepted the mat, holding it over my head to show that I was unworthy of such a tremendous gift. "From the bottom of my heart, I thank you. I wish to name this mat _molimau_ ("testimony"), for it will always serve as testimony and evidence that I am truly a member of Falealupo village."

The next morning, as I sat on the beach by our _fale_ watching the sunrise, I felt great pleasure at what had occurred the previous night, not only because I had deflected a _matai_ title that would have frustrated my ethnobotanical work, but also because I felt honored to be considered a full member of the village community. I was proud to be an untitled man in Falealupo. I felt as though I had truly returned.

6 cm.

Atuna racemosa

CHAPTER 5

Correspondence

I have escaped everything that is artificial, conventional, customary. I am entering into the truth, into nature.

Paul Gauguin, *Noa Noa*

I pulled the slide from the envelope and peered at it against the light outside our *fale*. What I saw was incredible; indeed, it beggared belief. Artist Michael Rothman had captured perfectly in a mural-sized painting the flight of a Samoan flying fox, and yet he had never left Manhattan. Working only from my published descriptions and a specimen skin he had found at the American Museum of Natural History, Rothman had managed to recreate the animal's majestic afternoon flight high above the rain forest canopy. I read his note again: "When I heard of your struggle to save the Samoan flying fox, I felt that I had to do something to help. I have worked three months on this painting, which is really more of a study. Your criticisms would be appreciated." I was astonished that Rothman had heard of my ongoing effort to protect flying foxes—other than a short piece in *Natural History* on the topic, I had published only a few articles in obscure journals.

"What are you going to do about this artist?" Barbara asked. "The least you can do is answer his letter."

"No, the passion that produced this painting deserves more than a letter. Let's invite him to Samoa."

Barbara was alarmed. It's one thing to sit in America and think about swaying palm trees and tropical breezes along the beach. It is another to confront the reality of Samoa—flies in your eyes, breakfasts of fish heads and taro, and nights sleeping on thin mats laid out on a crushed coral floor. Dr. John had lasted only three days—and that was in Apia. Because of the extreme remoteness of Savaii and our desire to live close to the people, we had decided not to bring any students or other foreigners into the village. And now we were considering an invitation to an artist we didn't even know, from that well-known bastion of outdoor ruggedness and pioneering spirit, Manhattan.

"Barbara, I think it's worth the risk. I've got a hunch about this guy. And if he doesn't work out, we can always put him back on the plane."

Six weeks later, the angry noise of an aircraft disturbed the tranquility of the morning long before we could see it approaching the Asau airstrip. I wondered what Rothman was thinking. Probably he was filled with the same foreboding all visitors feel when catching their first glimpse of the tiny airstrip extending into the lagoon. I sat on the bumper of our four-wheel drive and watched the cloud of dust settle after the plane landed. The Twin Otter taxied up to my vehicle.

The pilot opened the passenger door over the wing. Excited as I was to meet Michael Rothman, my attention was immediately drawn to a small bundle under his arm. Rothman had left Manhattan on Sunday, stopping en route only to change planes. Yet somehow on his thirty-hour flight, he had carried with him a pristine copy of the Sunday *New York Times*. Sizing up my interest immediately, Rothman pronounced in a seasoned Brooklyn accent: "I brought you a present."

During the drive back to Falealupo I tried to make polite conversation, but all I could think about was my unexpected windfall. The *New York Times!* How could I make the pleasure last? By the time we had driven through the Falealupo forest and pulled up along the white sand in front of our *fale,* it was clear to me that rationing was required: each day would merit a single page read in its entirety. Every night I retired to my mosquito net with a headlamp. Announcements by European ministers, book reviews by Norman Mailer, ads for fur sales at Macy's, NYSE quotes—all were consumed on their appropriate day. And by holding multipage splurges to a minimum, I was able to make the paper last for a long time. Soon, though, I was approached by a delegation of curious villagers. What was I doing each night in the darkened *fale* with the headlamp? Praying? Indulging in some sort of strange *palagi* nocturnal ritual?

"Reading," I replied.

"Oh," they said in disbelief.

But the pages of the *New York Times* had more uses than one, as Michael Rothman soon learned to his horror. "That's the toilet?" Michael asked with incredulity.

Barbara's face fell. Knowing adjustment to Samoan sanitary facilities to be second in difficulty only to adjustment to Samoan food for most *palagi*, Barbara had disinfected the little tin shed with bleach and Lysol. The sturdy concrete toilet inside, invented by the Peace Corps, offered a secure foundation that few recent visitors to Samoa could imagine. Formerly the toilets were rickety sheds built on stilts above the sea. Indeed, the Samoan name for toilet, *fale uila,* or "electricity house," refers to the wooden crates originally used to ship generators to the islands. Walking across precariously positioned rocks to these frail structures during high tide or storm tested the resolve of even the most determined patron. But even in good weather and at low tide, the village pigs would run toward the *fale uila* at the first sign of an approaching user. Seeing the faces of five or six hungry pigs framed in the toilet hole, more than one *palagi* has reconsidered their visit. Nothing like playing to an enthusiastic audience.

Of necessity some *fale uila* were built inland, but they too were possessed of their own peculiar terrors. During my first sojourn in Samoa, my missionary companion Kevin Sommer became severely constipated in Samata village, since every time he entered the darkened *fale uila,* his flashlight revealed large centipedes and a spider "the size of a baseball glove." Finally Kevin seized upon a desperate solution—he stopped taking his flashlight to the toilet. "Even the worst things I could imagine, sitting there in the dark," Kevin explained, "were not as terrible as the reality my flashlight would have revealed."

Another difficulty *palagi* faced with *fale uila* was more cultural than biological. Most Americans and Europeans see toilet activities as something best conducted in private. But the original crates that became *fale uila* were too small to accommodate Samoan standards of sociability—Samoans hate to be alone in nearly any situation. So it wasn't long before some enterprising chief built himself a two-seater. Soon, three- and four-seaters appeared, until finally, the mother of all *fale uila* was constructed in the village of Nofoali'i, on the island of Upolu: twenty-eight seats, fourteen on a side.

So while sympathetic to Michael's plight, I realized that he had it easy compared with the old days before the Peace Corps replaced all of the old *fale uila* with concrete water-seal toilets.

Otherwise, Michael made a relatively smooth adjustment to Samoa. Tall and lanky, with brown hair and a very distinct New York accent, Michael had an affable grace that endeared him to the Samoans, particularly Pela, who became quite maternal in her attitude toward him. Many mornings I found her sitting next to Michael, lecturing him in Samoan about what he should do with his life and marriage, while Michael sat grinning, not comprehending a single word. Pela even moved Michael into her *palagi*-style house inland from our *fale*.

"Koki, you and your family are like Samoans," Pela explained. "But Michael isn't used to our ways. It's not right to have him sleep on mats in your *fale*."

She arranged a room for Michael with a chair, a small table for his pens and paints, and a bed with an overstuffed kapok mattress. Michael strung red nylon cords across the room to suspend plants for drawing. In front of the table he placed the portable easel that he had brought with him from New York.

After Michael had a few days to settle in, we climbed the cinder cone to look for the creatures that he had so lovingly painted sight unseen. It took us about forty-five minutes to scramble to the top, making our way through vines and woody lianas. We sat in a small clearing and looked out over the Falealupo forest.

"Where are the bats?" Michael asked.

If we were quiet, I said, we might see the breeding pair. Soon, we saw soaring through the trees a large pair of black wings.

"Wow!" exclaimed Michael, his camera in hand.

The flying fox began to angle downward, and glided around the prow of the hill until it disappeared. I told Michael that he had frightened it.

We waited in silence for nearly an hour, binoculars in hand, but the flying fox did not reappear. The sun was perceptibly lower in the late afternoon sky when Michael whispered to me, "I've got an idea—I'll call to it."

Since we were leaving anyway, I had no objection. Michael stood, and impersonated a hog caller. "Soo-ee!" Then he called in a sweet singsong voice for the flying fox using its Samoan name, "Oh, *pe'a vao*. Come here, *pe'a vao*." A huge Samoan flying fox came soaring from the edge of the hill, flying straight toward us. Michael was ecstatic and began flapping his arms. "That's a good boy, *pe'a vao*. Come on closer, *pe'a vao*." The flying fox did exactly that, slowly flying to within twelve feet of us.

In my previous fieldwork I had attempted to remain as unnoticeable and as quiet as possible in my observations of these animals. Having been driven nearly to extinction by commercial hunters, any surviving members of the species must, I presumed, be extremely wary of human beings. And yet here was Michael, flapping his arms and singing to the bat. The flying fox slowly glided over our heads and then disappeared past the trees above us.

"Well, he might have been curious about us," I conceded, "but I'm sure he's gone for good now."

We waited quietly another half hour, but the flying fox did not reappear. Finally, as we stood to leave, Michael started calling again.

"Soo-ee!" Michael yelled. "Come on back, *pe'a vao.*"

A moment later, the bat came gliding back into view. The low angle of the sun made his wings almost translucent above the forest canopy. And so it went for the next hour and a half, with the bat disappearing and Michael calling it back into view.

As we descended the hillside in the cool of the evening, I felt chastened. An artist who had spent less than a week in Samoa had reminded me of a truth that I once knew as a child: a kind of magic still exists in the world. In our rush to label and analyze things we often forget this, and thereby lose touch with the very essence of nature. Children, poets, and a few scientists of genius still retain sight of this central mystery—that of all possible universes, we live in the most beautiful, the most simple, and yet sometimes the most counterintuitive. The voice within us can temper the urge to reduce the beauty of life to its bare components, and instead teach us of a deeper truth.

There is more than magic or even beauty to the presence of flying foxes in the Falealupo rain forest, however. They play an important role in the ecology of the island. During my doctoral studies I discovered that *Pteropus samoensis,* one of the two species of flying foxes in Samoa, pollinates the bright red flowers of the forest vine *Freycinetia reineckei.* The more I observed the flying foxes, the longer grew my list of plants that require it for pollination, and I began to suspect that I had stumbled on the major pollinator of the Samoan rain forest. Flying foxes also disperse seeds and fruits too large for birds to carry. Unlike the other Samoan species, the white-necked *Pteropus tonganus, Pteropus samoensis* forages mainly during the day, so it is easy to observe which plants it chooses to feed on.

Beginning in 1979, I began to notice a significant decline in the flying fox population. Many Samoans familiar with the forests corroborated my concerns, telling me of their own alarm at the sudden decrease in flying fox numbers throughout Samoa. Careful investigation revealed the existence of trade in dead flying foxes. Commercial hunters, armed with semiautomatic shotguns, were killing hundreds of flying foxes, loading them into ice-filled coolers, and air-freighting them to Guam, a U.S. territory, where they were sold for prices as high as thirty-five dollars each as luxury food items. The trade accelerated in the 1980s, with over eighteen thousand flying foxes legally exported from Samoa within a three-year period. Other information pointed to an illegal trade of at least that magnitude.

The impact on flying fox populations in both American and Western Samoa was disastrous. The white-necked flying foxes had declined dramatically in numbers, but could probably survive the onslaught if the hunting could be stopped. But I began to fear for the very survival of the endemic Samoan flying foxes. A biologist at the U.S. Fish and Wildlife Service encouraged me to petition to have the Samoan flying fox population of American Samoa listed as an endangered species. The suggestion to solicit the help of the U.S. Fish and Wildlife Service seemed a good one. As a boy, I spent a season in the far reaches of Saskatchewan, where my father, a U.S. Fish and Wildlife conservation officer, banded geese to determine their migratory paths. As I grew older, my mother, a fisheries biologist, eventually became an administrator for the Service for all of the Western states. Under her direction, the Service took a determined stance against poaching. Once she received a call from an elected Wyoming official who had been apprehended shooting bald eagles from a helicopter. He told her that it would be in her interest to quash the investigation. My mother ignored the threat and vigorously pursued criminal action against the official, even though he was nominated a few weeks later by Richard Nixon to become Secretary of the Interior. In short, I had grown up with an image of the Fish and Wildlife Service as a courageous protector of biodiversity. But unbeknownst to me, the philosophy of the agency had changed.

In the 1980s, the Endangered Species Act drew increasing criticism because of its potential interference with business development. Although the U.S. Congress refused to revoke the act, pressure was placed on the executive branch, including the Fish and Wildlife Service, to reduce its enforcement. Numerous petitions to place species on the endangered species list were subsequently ignored, lost, or otherwise not acted upon. Blissfully unaware of these political changes within the Fish and Wildlife Service, I filed a formal petition to place the Samoan flying fox, *Pteropus samoensis,* on the endangered species list. The first inkling of trouble came in a terse telephone call before I left for Samoa.

"Mr. Cox, this is the U.S. Fish and Wildlife Service. There seem to be some problems with your petition to place *Pteropus samoensis* on the endangered species list. It appears that this species doesn't exist."

"Oh?"

"Yes, there is only one species of flying fox in Samoa, *Pteropus tonganus. Pteropus samoensis* appears to be merely a name resulting from confusion in the literature. A team of our biologists working in Samoa found no evidence for their existence, and as recently as last year an English biologist during a layover at the Pago Pago airport could not sight a single one."

"My work on flying fox pollination is supported by extensive photographs and by my collections at Harvard University," I replied. "If anyone has any questions concerning the existence of *Pteropus samoensis,* they need only consult the specimens. They are clearly a distinct species from *Pteropus tonganus* and differ dramatically in size and behavior."

"Well, you are only a botanist, aren't you?" the official asked.

"Yes. Why is that important to my petition?"

"It's just that we have a lot of problems with your petition, problems a trained zoologist would have avoided. In your petition you state that *Pteropus samoensis* forages in the daytime. This of course is impossible since bats fly at night. And in your petition you state that *Pteropus samoensis* is large, with a wingspan approaching two meters—such a size is ridiculous."

"I know there are two species of flying fox in Samoa, and I have photographs of both species to prove it. As for wingspan, the size is an estimate I made from observing an extremely large individual through binoculars. I assure you that they are very large. Instead of trying to dispute my observations, why don't you protect the flying foxes? Every day more flying foxes are being killed."

"Mr. Cox," came the icy reply. "We get petitions from people all the time—housewives, animal lovers, kooks—and we just don't have the time or resources to investigate those that aren't credible. You want us to protect a flying fox that we don't believe exists, and your petition contains information that we have a hard time accepting as factual."

"I'd be pleased to send you reprints of my scientific articles on the flying foxes, but I don't believe it's my credibility as a scientist that is on trial here."

The official muttered something noncommittal, and rang off.

Before we left for Samoa, I arranged to meet the official there. I was confident that once I showed him the flying foxes, the need to protect them would become clear. When the species was first described in 1848, they were so numerous that their very smell permeated the Samoan rain forest. "Their spectacular appearance is one of the characteristics in the wild and varied scenery," wrote naturalist Titian Peale of the U.S. Exploring Expedition in 1848. "Their strong odour taints the atmosphere of the otherwise fragrant forests, and will always be remembered by persons who have visited the interesting regions inhabited by these animals."[47] Peale also observed both the distinctive diurnal behavior and the marked territoriality that I had witnessed: "It is the least gregarious, and most diurnal, in habits of any of the genus which we saw; they are frequently abroad at noon-day."[48] K. Anderson of the British Museum agreed with this behavioral description, writing in 1912 that *Pteropus samoensis* "may be frequently seen even at midday flying high in the air with a slow and regular flap of the wing, not unlike a small heron, and occasional short intervals of sailing."[49]

I was confident that if I could show the Fish and Wildlife official the species in nature, all misunderstandings would be immediately cleared up. Michael Rothman and I caught the ferry to Upolu to meet him, but waited at the Apia hotel in vain.

"You are sure that no one from the U.S. Fish and Wildlife Service has checked in or even made a reservation?" I asked the hotel receptionist.

"No, I'm sorry, no one."

We checked with Dan and Kathy Betham, whose address and telephone number I had left with the Fish and Wildlife Service. There had been no attempt to contact us. We checked all of the hotels and government offices and found no sign.

"It looks like they've stood us up," Michael observed.

"Well, let's send them a telegram, and take a plane to American Samoa. Maybe they're over there."

Later that afternoon, Michael and I sat on a rocky ledge high above Afono village in American Samoa. The climb up the knife-edge ridge had been conducted in frustrated silence. Our search of the guest registers in several of the hotels had finally revealed the visit of a Fish and Wildlife team the week before. Descriptions from the hotel clerks confirmed our worst fears. Biologists from the Honolulu regional office had apparently come to Samoa, but for some inexplicable reason had avoided us. We looked down the cliffs to the village, the white sand beach, and the vast blue sea beyond. Michael sat, binoculars in hand, with his feet dangling over the edge. In a period of a few weeks, he had become expert in spotting flying foxes.

"What do you see?" I asked.

"Oh, there's a few _Pteropus tonganus_ in the tree down there, starting to wake up and quarrel. But I don't see any _Pteropus samoensis_."

Suddenly from around the edge of the ridge, a large _Pteropus samoensis_ glided like a black Stealth fighter. Its path circled silently to the cliff below us, where it hung motionless, riding the updraft like an invisible elevator to the level of our ledge. Suspended in air, the flying fox, with its deep brown eyes and blond forehead, seemed almost within touching distance. I could clearly see the finger bones that provided the structural support for its wings, and, more ominously, bullet holes in the wings. Michael gasped as the flying fox tilted its head, looking at each of us in turn. Its eyes were not sad, but piercing—as if the flying fox considered us inferior creatures trespassing on its territory. Apparently satisfied with its inspection, it pulled its wings back like a jet fighter and plummeted nearly fifteen hundred feet, just barely pulling out of its dive above the forest canopy.

Michael was speechless. I muttered under my breath, "Careful, babe, you're not even supposed to exist."

On our way back to Savaii, we visited Kathy Betham in Apia to see if we had received any mail. Although Pago Pago had been sunny, it was raining in Apia. We found Kathy at her small shop in the open-air market, where she produced an envelope from the U.S. Fish and Wildlife Service. "Dear Mr. Cox," the letter began, "We are sorry we missed you in Apia. We were very busy . . ."

I read on. The Fish and Wildlife Service claimed that in a period of a few days they had seen hundreds of Samoan flying foxes and were convinced that they were in no danger. As a result, they were going to decline my petition to list the flying foxes as an endangered species. I showed the letter to Michael.

"This is incredible," Michael said. "They're either lying, or they've confused the two species."

To an untrained observer, the white-necked flying fox *(Pteropus tonganus)* and the Samoan flying fox *(Pteropus samoensis)* look similar, particularly when viewed in flight from below. The white-necked flying fox is far more common than the Samoan flying fox, living in colonies of up to five or six hundred individuals.

"Well, in either case, I'm afraid this is a death sentence for the Samoan flying fox. Unless the commercial hunting stops, they'll go extinct."

When Michael and I returned to Falealupo, Lamositele told us more bad news. The village had received final notice from the government that it would have to come up with funds to build a new school, with an estimated cost of $65,000 U.S. For villagers whose per capita income is less than $100,[50] raising such an amount was scarcely possible. The village had appealed to the government asking them to certify its current school as acceptable, but the government was adamant. If a new school was not constructed in the next year, the village children would be denied entrance into the Western Samoan educational system.

Through one of those mysterious coincidences that frequently happen in the Third World, the order to build the school was soon followed by yet another offer from the logging company to harvest the thirty-thousand-acre Falealupo forest. For ten years running the village had resolutely turned aside all requests to log the forest. This time the loggers offered $65,000 U.S. for the rights to log it.

"It's awful," Seumanutafa told me the next day, sitting in his *fale*. "We either sacrifice our children's future or we destroy our forest," he said sadly. "We have no other alternative. And you know how long we've protected our forest. There isn't a single chief in the village who wants to see it destroyed."

"Seumanutafa, you know how precious this forest is," I said. "There must be some alternative. Please ask the chiefs' council to deny the loggers' request."

"I'll voice your sentiments, Koki. But we can't abandon our children. They are the future of our village." The village elders knew that the traditional way of life was changing, and that most of their children would need an education to compete in the modern world—subsistence skills such as reef foraging or tending taro patches simply would not be sufficient for integration into the larger political economy of the South Pacific.

That next week Pela's family held a special *fa'amavaega,* or farewell celebration, at our *fale,* since Michael was returning to New York and we were leaving the following day for several months of research work in Australia before returning to Samoa.

The next day, as we drove out of the village, I wondered if the Falealupo forest would survive for long. Nearly all of the other rain forests of Samoa had disappeared into the jaws of the sawmill—could the villagers withstand the pressure to cut it down?

Our small house in Templestowe, a suburb of Melbourne, overlooked a park where Paul Matthew and I spent hours sailing a radio-controlled boat I bought him. The children rapidly adjusted to the Australian school system and made many friends. My colleagues at the University of Melbourne, where I had been awarded a research fellowship, treated me kindly, and my research topic—the pollination of seagrasses and other plants that flower underwater—produced fascinating results. The laboratories were superbly equipped, and my chief collaborator, Professor R. Bruce Knox, was a world leader in the field of pollination biology. Everything was wonderful on the surface, yet I was desperately depressed.

I dreaded going to sleep, for it was then that the nightmares came. I would see my mother and be overjoyed to realize that she was still alive. But as I embraced her, I realized that she was still filled with cancer. I'd awake in a sweat and try not to drift off, lest the dream come again.

My days were emotionally difficult as well. I tried to analyze pollination data, but whenever I closed my eyes for a moment I envisioned instead the carcasses of dead flying foxes or heard the rumble of bulldozers destroying the Samoan rain forest. I saw the Falealupo villagers scratching out a living from burnt ground much like the remains of the logged forest I had seen in Tufutafoe. Although I tried to maintain a facade of cheerfulness, I felt scarcely able to make it through each day.

After several weeks I confided my depression to Barbara, but I rejected her suggestion of counseling. My despair was not symptomatic of neurosis, but rather anguish at my impotency to change the course of events. My mother had died an agonizing death despite every attempt to save her. And the entire Samoan ecological triangle—the Samoan culture, the flying foxes, and the rain forests—was being eroded as well, and its demise appeared inevitable. I pursued my petition to have the Samoan flying fox declared an endangered species, but the Fish and Wildlife Service continued to respond with attacks on my professional competence. The dreams of my mother continued, and each day I felt more morose than the last.

"Paul," Barbara finally said, "I'm really worried about you. You can't fight the entire government alone. Why don't you call Bill Rainey and Dixie Pierson? Perhaps they can help."

I thought for several days about Barbara's suggestion. In a way, Samoan flying foxes had led to our acquaintance with Bill and Dixie, zoology graduate students at the University of California at Berkeley in 1981–1983 when I was a Miller Research Fellow there. Bill attended a seminar I presented on my work in Samoa and was struck by my description of flying foxes as pollinators. That same day Dixie, who was attending some scientific meetings at Cornell, acquired from a rare book seller an original print of the Samoan flying fox, *Pteropus samoensis,* published in the Atlas of the U.S. Exploring Expedition. Dixie's doctoral dissertation on New Zealand sheath-tailed bats had been brilliant, and Bill had achieved international distinction as an endangered species biologist.

When I invited Bill and Dixie to come to Samoa, they told me that they were unfortunately booked up for a year with fieldwork. Bill did give me some comfort, however, in letting me know that I was not alone in my difficulties with the Fish and Wildlife Service. Dr. Tom Lemke's petition to place the Guamanian sheath-tailed bat on the endangered species list was denied because Fish and Wildlife biologists couldn't find any living individuals—in a maneuver reminiscent of *Catch-22,* they turned down the petition on the basis of 'insufficient data.' The Service had also turned down a number of bird listings throughout the Pacific.

When I told Barbara that Bill and Dixie couldn't come to Samoa until next year, she came back with another suggestion: Merlin Tuttle, who had seemed very interested in my work at Harvard.

When I finally got in touch with Merlin Tuttle, founder of Bat Conservation International, he was very positive. "I've been wanting to photograph those Samoan flying foxes for my new book ever since you told me about them. In fact, a Milwaukee businessman,

Verne Read, has agreed to finance a trip to Samoa, particularly if his wife can come along."

"How old are they?"

"Well, they're mid- to late sixties, but honestly, Paul, you'll like these people," Merlin reassured me.

"Merlin, I'm sure I would *like* them, but having a Milwaukee businessman and his wife with us during fieldwork in Samoa is a different story. It isn't Club Med there. It's hot, humid, and a very different culture."

"They really want to come, Paul. I sort of owe it to them— they're the ones who bankrolled Bat Conservation International in the first place."

I flew to Fiji several weeks later. At a breakfast of fresh papaya on the veranda of the Tanoa hotel, I awaited my visitors. Merlin entered, impeccably dressed in sharply pressed khakis and a blue blazer. With him walked an attractive couple dressed in tennis out-fits and equipped with rackets.

"Good to see you made it, Paul. Please meet Verne and Marion Read."

Verne wiped his hand on his sweatband and clasped my hand firmly. Marion smiled and extended her hand.

"Sorry we're a bit late," Verne volunteered as we sat down. "Marion and I just finished our morning set."

"Do you play a lot of tennis?" I asked.

"Oh, now and then, when we have a chance," Verne answered.

"Don't you believe it," Merlin laughed heartily. "They've just returned from the Senior National Championships, where Marion took first place."

Marion smiled demurely.

"Tell me, is this your first trip to the tropics?" I asked.

"Well, Marion and I spent a good part of last year in Costa Rica and Panama doing research with Merlin," Verne said.

"And don't forget our expedition in Thailand the year before to find the world's tiniest bats," Merlin said.

"So this is your third trip to the tropics?" I asked, still surprised that, far from being the overweight and underexercised Milwaukee couple I had expected, Verne and Marion seemed to be in far better physical condition than I was.

"We've had a few others," Marion said modestly. "Last spring we went to Africa to climb Mount Kilimanjaro with our children."

"You climb, too?" I asked.

"Verne is on the board of the American Alpine Club," Merlin responded, relishing my discomfort. "Last year they hiked into the base camp at Everest and made a first ascent up an unclimbed peak."

After visiting, we prepared for the flight to Apia, where we spent a night at Aggie Grey's Hotel before catching a connecting flight to American Samoa. The next day Merlin and I struggled to keep up with Verne and Marion as we hiked along the steep path above Pago Pago to the rain forest ridge where we hoped to photograph flying foxes. "Can't they slow down just a bit?" I asked Merlin. Soon we saw several Samoan flying foxes below the ridge, and Merlin's cameras whirred. Merlin and the Reads were particularly struck by the vulnerability of these diurnal bats to hunters, and became further concerned when we stumbled one day across a commercial bat hunter, who showed us a freezer full of dead animals.

Although the prospect for the flying foxes was bleak, I was grateful for Merlin's corroboration of my scientific account. And a fast friendship was developing between myself and the Reads. With their considerable financial and social success, both Verne and Marion had decided to devote their lives to conservation. After seeing firsthand the plight of the Samoan flying foxes, Verne asked what the best way of protecting them would be.

"Since American Samoa is a U.S. territory, the Endangered Species Act would apply here. But the ultimate federal protection for U.S. lands is a national park, which protects not only endangered species, but entire ecosystems," I replied.

"So why don't we make a national park here?" Verne asked innocently.

Oh, boy. "Verne, it isn't that easy," I patiently explained. "There are only forty-nine national parks on U.S. lands, and each one requires a special act of Congress. I'm not even sure that Congress knows where Pago Pago is. I suspect the local chiefs here in American Samoa would be pleased with some sort of protected status for their forests. But politically, any initiative for a national park would have to begin with American Samoa's Governor, and . . ."

Verne picked up the phone. Within an hour, we were in Governor A. P. Lutali's office. With Merlin and Verne in strong support, I broached the idea of establishing a U.S. national park in Samoa. Eni Faleomavaega, the lieutenant governor, also spoke in favor of the concept.

"Why not?" Governor Lutali said.

I was stunned at how easy everything seemed. Eni Faleomavaega, himself a Samoan chief and a law graduate of Berkeley, pointed out a few obstacles, however.

"Customary land can be neither bought nor sold under the American Samoan constitution, and there is no federal land here on which to base a park. But perhaps it could be leased. . . . The American Samoan Constitution specifically allows for the establishment of leaseholds on customary lands for up to fifty-five years. I used to work on the House Interior Committee for John Siberling, but I'm not sure if Congress would buy it."

"I know John, as well as Bob Kasten in the Senate. Let me see what I can do with Congress," Verne volunteered.

I offered to organize a series of meetings in the villages with the local chiefs, and Merlin agreed to coordinate media coverage.

I returned to my family in Melbourne with renewed hope. But when I opened the confidential letter from Dr. Gordon Cragg at the National Cancer Institute, I realized that far more was at stake than the flying foxes and the Samoan rain forest. In addition to screening my plant samples for anticancer activity, the National Cancer Institute had also asked to screen them for activity against the AIDS virus, HIV-1. Dr. Cragg's letter detailed the results of the latter investigation: *Homalanthus nutans,* the tree that healer Epenesa Mauigoa told me was useful against hepatitis, was extraordinarily active against the AIDS virus. Given concern about prematurely raising the hopes of AIDS patients, news of the discovery would be restricted to a small group of researchers. It would be very useful for me to collect more samples of the plant as well as healer preparations for further analysis. Could I provide them? I was thrilled to assist.

Our remaining time in Melbourne passed quickly. For Christmas, Barbara presented me with an entire case of fresh lychees, a tropical fruit commercially unobtainable in the continental United States. Since Christmas in Australia signals the advent of summer vacation for schoolchildren, we flew back to Samoa after the holiday, arriving in Falealupo shortly before the New Year. I immediately asked Seumanutafa Siaosi about the status of the forest and the logging offer. He told me that the village had petitioned the government to certify the local Catholic school as satisfying the requirement for an adequate school. The villagers treasured the rain forest, Seumanutafa assured me, and would take all the steps they could to protect it. For a while the forest seemed safe, but everything ultimately depended on the government's response.

January is often stormy in Samoa, but even though severe storms occasionally kept me out of the forest, the inclement weather provided additional opportunities to interview Pela and the other villagers. I was particularly interested in recording with notebook and video the techniques Falealupo healers used to prepare the plant with anti-AIDS activity, *Homalanthus nutans.* My daily interviews with Pela were supplemented by discussions with other Falealupo healers, including Lemau Seumanutafa, Lala'ai Silia, and Mariana, Pela's sister-in-law.

Mariana's house was located on a small hill about a hundred meters behind Pela's house. In front of it, growing from a crack in the lava, was a relatively rare tree called *ifiifi (Atuna racemosa)*, with beautiful egg-shaped fruits about the size of a fist. Mariana showed me how she used the fruits to make a special massage oil for herself and Pela. Samples of the fruits that I had sent to Lars Bohlin's laboratory in Uppsala, Sweden, demonstrated potent anti-inflammatory activity. This beautiful tree, with its simple green leaves and gracious brown fruits, facilitates healing in Mariana's and Pela's hands. It is of course anthropomorphic to assign moral values to plant species, but were such a judgment to be made, the *ifiifi* tree should certainly be declared good. "Trees especially seem to bespeak a generosity of spirit," Annie Dillard writes. "I suspect that the real moral thinkers end up, wherever they may start, in botany."[51]

I also intensified my study of a plant that Polynesians explicitly assign a moral value to, a plant they believe can promote social tranquility and even facilitate spiritual transcendence: *Piper methysticum,* or kava. But I required help to understand the cultural significance of kava. The rhetoric associated with it was so complex that I began to worry that there might be some things Polynesian that can never be understood by a *palagi,* no matter how well versed in the language or culture. The soul of a Polynesian "is not revealed immediately," Paul Gauguin wrote in his Tahitian journal. "It requires much patience and study to obtain a grasp of it. And even when you believe that you know it to the very bottom, it suddenly disconcerts you by its unforseen jumps . . . it is enigma itself, or rather an infinite series of enigmas."[52] I required a teacher to help me master the enigma of kava.

Lamositele's brother, Lilo Manuele, provided me with the needed instruction. Lilo Manuele adores plants, and his kava plantation is the envy of the village. He is proud of the orator's title, "Lilo," that he inherited from his father, and excels at the part of the kava ceremony called the *folafola 'ava.* At this stage of the ceremony, the orator representing the visiting party removes his shirt, slides forward on the mat, and respectfully receives the village's gift of kava to the visitors. When Lilo Manuele recites the *folafola 'ava,* each word is carefully enunciated in a high ringing voice: " . . . *ae o le ā te'a atu le lupesina lālelei lenei i le tanoa, e lele mālie atu, ae afai e taunu'u i le aumaga o tofiga, o le ā latou lu'ilu'i ai e fai ma agatonu o lo tatou taeao fesilafa'i"* (". . . and this beautiful kava root is dedicated to the kava bowl and will gently float to the untitled men, who will lovingly prepare it to become a symbol of our fellowship and love this morning"). After all of the speeches are completed, the numbing beverage made from the kava root is ceremonially consumed by all.

Just as Mass was long recited in Latin, so the kava ceremony is conducted in formal Samoan respect language. As such, the discourse of the ceremony is not easily followed even by those fluent in the common language of Samoa. In eloquence and sheer power of language, the kava ceremony is unequaled by any communal expression I know of in Western culture.

> Our meeting is as sacred as the tips of two clouds passing
> in the sky. Our meeting is as the mating of sea turtles,
> silent, motionless, and sacred. Our meeting is as sacred as
> the first dew, as sacred as the first ray of light that filled
> the newly created earth. Our meeting is as sacred as the
> meeting of the sea and mountains, who looked upon
> the sun and asked why it wept. The very mountains and
> the waves of the sea are moved at our meeting on this
> morning—this morning that above all mornings is clear
> and calm, and sacred. We meet in the midst of sacredness.
> The sea is sacred, the earth is sacred, our meeting ground is
> sacred, our meeting house is sacred, and it is with fear and
> trembling that we speak of the sacredness and power of
> those whom we now address . . .[53]

Samoa really has three languages: the common language spoken on an everyday basis, the respect language spoken to honor others, and the language spoken by orators during kava ceremonies, such as that of Lilo Manuele's *folafola 'ava*. Young Samoans, even those who have learned some respect terms, don't have the slightest idea what is going on when the chiefs start using kava rhetoric. A kava ceremony can resemble a scholarly symposium replete with footnotes, literary allusions, and veiled intellectual jousting. But unlike academic discourse, in Samoa kava metaphors frequently revolve around a topic that once influenced the language of the English Norman court: fowling. And Lilo Manuele played a key role in helping me understand fowling allusions in kava speeches.

In ancient times, the capture of terns and fruit pigeons was a royal sport in Samoa. Terns were captured with long-handled nets as they flew up cliffs. Success in ensnaring pigeons, however, was determined by the quality of the hunter's live decoys, which were trained to obey instructions communicated by tugs on strings attached to their feet. A decoy was tied to a stone platform near a blind, where the hunter, with his head camouflaged by foliage, would sit on a small stool. The decoy would be sent aloft, and through tugs on the string, directed to fly one way or another. If a wild pigeon came close to the blind, the hunter would raise his long-handled net to deftly seize it.

"Samoans take infinite trouble in training their decoy birds, and value them very highly on their education being satisfactorily finished," William Churchward, the British Consul to Samoa, recorded in 1887.

> Indeed, they become so attached to them as to never let them out of their sight, feeding them in person, and even taking their pets with them when abroad. The reputation of a superior bird will spread all over the island, and become the talk of sporting men in every district.[54]

Fowling, an expression of high social status, was often played for high stakes: the entire village of Tafua, Savaii, is said to have been won during a fowling contest. This ancient sport, nearly forgotten in modern Samoa, remains alive in kava rhetoric:

Ua tatou 'oa'oa i faleseu.	*"We have joy in our fowling blind." (Expresses joy at a chiefly gathering.)*
Ua liligo le fogātia.	*"The hunting ground is silent." (Indicates the sacredness of an occasion, such as a kava ceremony.)*
Ua vae lupe maua ae lē vae lupe sa'ā.	*"We have the leg of the captured pigeon, rather than the one that fled." (Indicates joy and prosperity.)*
O le ā sosopo le manu vale i le fogātia.	*"The foolish bird will fly over the hunting ground." (Indicates humility before speaking.)*
Ua malu maunu le fogātia.	*"The fowling ground is dignified by the decoy." (Expresses respect.)*
'Aiga le fale seu.	*"The fowling blind is occupied." (Indicates a dignified and joyous occasion.)*
Ua atagia taga tafili.	*"The movement of the decoy is known." (A concealed motive is now apparent.)*

Since fowling is no longer practiced in Samoa, the ability to use such proverbs knowledgeably and to discuss pigeon snaring techniques is almost strictly limited to chiefs. The complexity of fowling metaphors would have been nearly impossible for me to understand without Lilo Manuele's help. And only recently, when an orator in Ta'ū told me the story of the first kava ceremony, did I begin to sense there was a deeper significance to this indigenous liturgy. In that story, Tagaloalagi, the creator of the world, sat facing the first man, Pava, while preparing the kava. The *alofi,* or space between the two parties in a kava ceremony, is sacred, and once the ceremony begins, it is forbidden to stand, walk, or intrude into the *alofi.* Yet during this first kava ceremony between god and man, Pava's son impinged on the sacred space.

"Forbid your son," Tagaloalagi commanded Pava. "He must not violate the sanctity of the *alofi.*"

Twice Pava forbade his son, but the impatient youth, oblivious to danger, ran between the two celebrants. Tagaloalagi seized the child and tore him limb from limb.

Pava, devastated, cried out at the death of his only son. Tagaloalagi looked up at Pava. "Your son died because he violated the *alofi.*" Tagaloalagi lifted the kava cup. "But since through transgression came death, through the sacred kava will he live again."

Tagaloalagi let a few drops of kava drip onto the boy's dismembered body. The boy came back to life on the spot, his body completely healed. "And the sacred kava will always serve as an *agatonu* between God and man," Tagaloalagi concluded. Overjoyed at the resurrection of his son, Pava began to clap. And then he and Tagaloalagi imbibed the sacred beverage.

Even today during kava ceremonies, participants clap their hands and dribble a few drops upon the mat before drinking. Struck with this allegory of the fall and resurrection of man, I began to tape-record the kava ceremonies I witnessed in Falealupo, with an eye to translating the kava rhetoric into English. But I never found a satisfactory English equivalent for *agatonu.* Reduced to its components, *aga,* or "behavior," and *tonu,* or "correct," the etymology seems simple enough. But *agatonu* is such a powerful word that even the translators of the Bible avoided it. Thus, although Samoans commonly refer to Christ as their *agatonu* with God, this word does not appear in the Samoan Bible, and a different term, *togiola* (literally, "paying for life") is used for "atonement."

Perhaps the translators were justified in their reluctance to use the word, but the meaning of *agatonu* was still puzzling to me. For example, an untitled man meeting a chief should, in deference, declare that he has no *agatonu* in relation to him. Kava is referred to as *agatonu.* And when a village and a traveling party meet for the first time, the orators speak of searching for an *agatonu*. Even Lilo Manuele was unable to help me translate the word *agatonu*. One night I asked an old and distinguished orator, Fetui, if he could explain the concept of *agatonu*.

"You have not asked an easy thing," he said.

"How about 'satisfy'?" one of the other orators present asked.

"No, that's not quite it," I responded, "for what does it mean to say to a chief that I don't have any satisfaction with him? How can I express *'agatonu'* in other Samoan words?"

The other chiefs fell silent. Finally one said *"fetaui"* ["correspond" or "fit"]—or *"fa'afetaui"* ["to make to correspond" or "to make to fit"].

"That's it," said Chief Fetui. "*Agatonu* means 'to make to correspond.' Have you ever noticed how people always drink kava in pairs? First the highest-ranking member of the traveling party is served, and next the highest chief of the village is served. And then kava is taken to the second-ranking member of the traveling party, and to the second-ranking member of the village. This continues until everybody has reached some correspondence with a village member at the appropriate level. When you say to a high-ranking chief 'I have no *agatonu* with you,' what you are really saying is 'I have no right to be in your presence because I have no proper correspondence with you, no equivalent rank.' Kava allows us to make correspondence with others."

"And is that why Samoans say Christ is their *agatonu* with God," I asked, "because He mediates that relationship?"

"Precisely," said Chief Fetui.

In the Polynesian world view, new people visiting a village are potentially very dangerous, filled with *mana*, or spiritual power, since they represent all of their extended kin, dead ancestors, and unborn descendants. It is only kava that allows a discharge of potentially dangerous spiritual power between the visitors and the villagers, by grounding each person with his or her equivalent in the village. Each meeting is "as sacred as the touching of two clouds in the sky," and just as potentially hazardous: if you do not achieve *agatonu*, you might be killed by lightning.

If the search for correspondence is the underlying theme of the kava ceremony, I wondered, who is my equivalent in Falealupo? Barbara had certainly bonded with Fa'asaina, and I began to suspect that my correspondent might be Lilo Manuele. And apparently he had similar thoughts. As we relaxed one evening, he asked, "Koki, do you think I am like you?"

This was an interesting question coming from someone who was obviously gifted in oratory and physically well endowed. Lilo

Manuele has a rugby player's build and a strong square jaw. When he speaks, he looks you in the eye with warmth.

"I'm neither as good looking nor as eloquent as you are, Lilo Manuele."

"You're such a flatterer, Koki. But I'm serious. I think I'm a lot like you. You love to learn about the old ways, to listen to the old chiefs, and to learn more about Samoan culture," he said. "So do I."

As the weeks and months passed, our children began to feel far more comfortable with both the culture and the language, aided in their efforts by Fa'asaina. One day, after lunch, Fa'asaina asked each of the children what they would like to be when they grew up. Emily said that she would like to be an astronaut. Paul, enamored of Lamositele's canoe and spear, said that he hoped to become a fisherman.

"And what do you want to be, Mary?" Fa'asaina asked.

"A pineapple."

The answer was quickly translated for the benefit of a few villagers visiting in our *fale*. Laughter immediately erupted. I asked Mary why she wanted to be a pineapple.

"Because pineapples are nice plants, they taste good, and make everybody happy."

Mary certainly was not joking about the taste. Though Samoan pineapples, ranging in size from footballs to small hand grenades, would never meet Dole Company standards, their taste is deliriously sweet, unlike anything we have ever purchased in a can or from the produce section of a supermarket.

So many things in Samoa were like that—without any analogue in Western culture, but absolutely delightful. I began to wonder if living in a simple *fale* in Falealupo and luxuriating in the company of the villagers was better suited to our family than the academic life in America and Australia that we had become accustomed to. Perhaps our family had achieved *agatonu*—correspondence—with an entire village.

After nearly a year in the village, it was with considerable sadness that we bid farewell to Pela and her family to return to my professorial responsibilities at BYU. Although I planned to visit Falealupo in the future, it was unclear when our whole family might have an opportunity to return. The morning we left, though, Seumanutafa told me something that left me shaken.

"Why is Daddy so upset?" Mary asked as we drove out of the village through the forest.

"He just learned that the government rejected the village's petition. The forest will be logged to pay for their school."

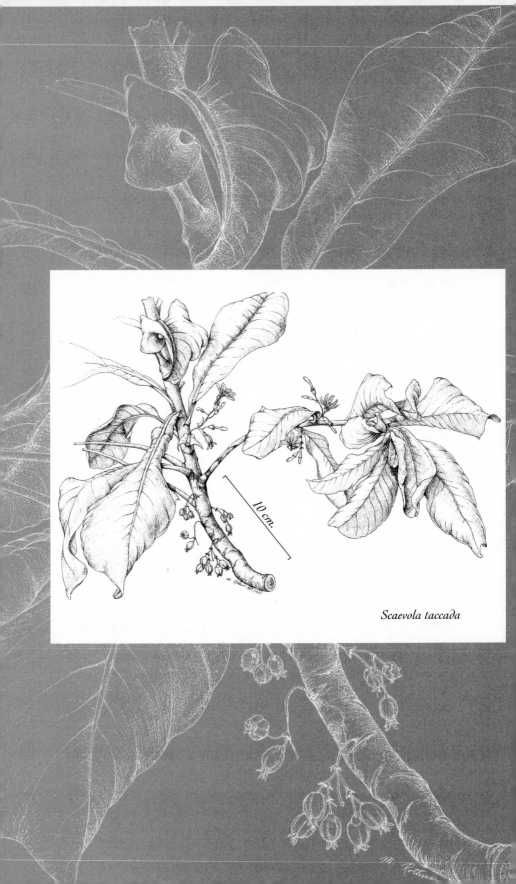

Scaevola taccada

10 cm.

CHAPTER **6**

Apprehension

*I have returned to my
beginning. I realize that if
through science I can seize
phenomena and enumerate
them, I cannot, for all that,
apprehend the world.*

Albert Camus,
The Myth of Sisyphus

The next July, as I waited at the exit from customs at
the Pago Pago airport in American Samoa, I feared
that Dixie might not be on the plane. Although her
husband Bill could not break free from his summer
work, Dixie had agreed to spend two weeks in the field with me
and my undergraduate research assistants, Ian Gurr, a Samoan student from Brigham Young University, Hawaii, and Tanya Vincent,
an ecology student from the University of Arizona. As was typical,
the airport was jammed with Samoans. Lacking a large movie theater, Pago Pago's chief entertainment was the airport, and since
everyone is somehow related to everyone else, there was always an
excuse to go out there and greet the incoming planes. The line of
people trickled out from the customs hall. Ministers of Samoan
churches wore oppressively hot suit coats and dark lavalavas. Several Samoan chiefs, inscrutable behind dark glasses, carried leather
satchels. Korean fishermen, returning from furloughs home, preceded several *palagi* schoolteachers, happy with knapsacks and
thongs. Toward the end of the procession were a few conservatively dressed Americans who disappeared rapidly into dark cars,
probably National Security Agency officers who operate the secret

submarine communications facility nearby. Two young Mormon missionaries, serious in white shirts and ties, followed, evoking in me feelings of nostalgia. The very last passengers were several befuddled Tongans waiting for the next day's plane to Nukualofa. But there was no Dixie.

Just as I was preparing to leave, I saw a baggage trolley coming toward me, stacked with two sixty-gallon Coleman coolers and a canvas bag of aluminum mist net poles on top. Behind it beamed Dixie. I'd forgotten that she and Bill always travel light. Dixie is blond and slender, with a tremendous intensity that reminds me in some ways of my own mother.

The next morning, Dixie, myself, and the two students were perched on a precarious ledge on a cliff nearly five hundred meters above Afono village, just over the knife-edge ridge towering above Pago Pago. Just as when Michael was there, the big Samoan flying fox with the blond head rode the updraft to see us, hovering suspended in air a few meters in front of us.

"He's fantastic," Dixie whispered. "But look at the bullet holes in his wings! It's a miracle he's still alive."

"Blondy," as Dixie christened him, looked us over for a minute or two, and then, apparently satisfied, dove down to the jungle canopy below.

We spent the next two weeks in American Samoa carefully recording the foraging patterns and food choices of the flying foxes, noting their preference for roosting near cliffs or large drop-offs. Our interest in flying fox conservation had garnered some local attention by this time. The territorial legislature had passed a measure banning all export of flying foxes from American Samoa. And in Pago Pago, a local conservation society, called 'O le Vao Matua ("The Rain Forest"), had been organized. The society soon began running ads in the local newspaper urging protection of flying foxes. Things seemed to be looking up for the cause of conservation in Samoa. Best of all, Dixie was captivated by the Samoan flying foxes, and agreed to return with Bill for an extended field season the following year.

After leaving Dixie in American Samoa, I traveled to Falealupo to inquire about the status of the forest. The village was still refusing all overtures from the logging company. Somehow, they believed, they would be able to reach a compromise with the government. I decided to use my National Science Foundation award to fund a major research effort on plant/animal interactions in Savaii the following year. That effort would be assisted by an important new member of our team.

Several months earlier, I had been invited by Umeå University in Sweden to be the external examiner of Thomas Elmqvist's doctoral dissertation. Elmqvist had written a brilliant dissertation on plant ecology, and it was without hesitation that I invited him to do post-doctoral studies with Bill, Dixie, and me in Samoa. His scientific training, not only in Sweden but also in Costa Rica, had prepared him well to investigate plant/animal interactions in the Samoan rain forest, particularly those involving flying foxes. His wife, Eva Pontén, was an M.D. who could assist me in my ethnobotanical studies, and their children were young enough to adapt quickly to Samoan life. The nature of Samoan society, with its emphasis on caring for the poor and elderly, was also clearly congruent with the ethics of Swedish social democracy.

In June 1988, I flew with the research team, which now included Bill Rainey as well as Dixie, Thomas, and Eva, to Pago Pago. They stayed on there a few days while I went ahead to make final arrangements for them in Falealupo.

It was wonderful to have the researchers with me in Falealupo. Even though it rained frequently, the scarcity of culinary water in the village was immediately felt by all. We all had to be very careful in our use of water for showers and washing clothes. But they didn't complain—instead of griping about inadequate water for bathing, the hard sleeping mats, or the strange food, they expressed gratitude to the Samoans for the courtesies showed to them.

I was eager to introduce Bill to the Falealupo villagers because I thought he could have fun helping me determine the biological identity of the most feared organism in Samoa, the *tanifa*. Was it a huge shark, a fish of tremendous size, or a killer whale? All I knew was that the Samoans feared *tanifa* above all sea-dwelling creatures. Press Samoans as to what type of organism the *tanifa* is, and they will invariably reply *"i'a,"* a Samoan term usually translated as "fish," but which includes all marine organisms. With Bill's encyclopedic knowledge of marine biology, perhaps I could solve the conundrum once and for all.

Villagers who don't give a second thought to lassoing sharks from dugout canoes, or shooting giant moray eels with crudely designed spears, become subdued when *tanifa* are discussed. All claim that the inherent ferocity of *tanifa* can lead to tragedies similar to what happened in Samatau village, on the island of Upolu. The Samatau men went out in dugout canoes to capture a *tanifa,* after having first prepared a fish trap made of sticks placed upright in the reef in a long, V-shaped pattern. To entice the *tanifa* into this fish fence, a live cow was tied at the apex of the trap. The plan was to find a *tanifa* in the open ocean, and then to drive it into the trap.

But someone was needed to stand behind the cow with a club to kill the *tanifa*. A visiting Fijian was prevailed upon to fill the crucial position. The fishermen, in a long line of dugout canoes, sighted a *tanifa* and began to chase it toward the trap by pounding on the sides of their canoes. The *tanifa* swam closer and closer to shore. But once inside the trap, instead of turning to flee seaward, the *tanifa* swam at high speed directly toward the shore. The cow went down, and blood filled the water. The *tanifa* circled back, its massive fins sticking out of the water, and then again sped toward shore, biting the Fijian completely in two before escaping through the apex of the trap.

Given the ferocity of *tanifa*, any Samoan who successfully captures one is assured lasting fame. By custom, a fisherman who catches a *tanifa* must not return to shore without first hoisting a shirt or lavalava on a paddle. Once the signal is sighted, word spreads like wildfire through the village. The village women unfurl their fine mats and stand in a line up and down the beach as if receiving royalty. Only then can the lucky fisherman bring his catch to shore. Since *tanifa* is considered sacred, the fisherman must deliver it to the chiefs.

The celebrity of fishermen who have caught *tanifa* is extraordinary. In Lotofaga, I met a man who introduced himself by simply saying, "I am Simi, catcher of *tanifa*." He said that during a fishing expedition, a *tanifa* had seized his bait, dragging the entire canoe out to sea. Slowly and patiently, Simi played the great *tanifa* with his hand line until the sun had nearly set. When he finally maneuvered the tanifa near his dugout, he beat it on the head with a club he kept for killing sharks, and lashed it with cord between the outrigger and the canoe. The *tanifa* exceeded the length of his canoe by two arm spans, he claimed, a size later corroborated by other villagers. That would mean the creature was fifteen feet long—longer than a Volkswagen van.

It had become dark, so Simi slept the night in his dugout, determined to paddle to shore the next morning. On waking, he noticed that his canoe was moving through the water. Using a paddle as a rudder, he guided his boat back to shore, propelled by the captive *tanifa*.

Lamositele related a similar story from Falealupo. While fishing in their dugout, a man and his young son drifted far out to sea, until the island was only a small speck on the horizon. They were about to return when the father saw a *tanifa* surface. "Son," he said to the boy, "it appears that God has blessed us with a _tanifa_. Is your heart pure? Should we pursue the _tanifa?_"

The boy nodded in the affirmative. The man looked gravely toward the _tanifa_ and began the ancient chant:

"_Afio mai, lau afioga._" ("Welcome, your Lordship.")

And just as the sharks and turtles come to the song of the children of Vaitogi village, the _tanifa_ responded to the fisherman's intonation. But it came with fury, seizing the bait and diving for the bottom of the sea, dragging the canoe with it. The boy and his father swam for their lives. They knew that they couldn't make it through the rocks and breakers in the dark, so they determined to stay afloat all night. At first light, they saw a small dugout canoe in the distance. Swimming to the canoe, they discovered that it was their own—attached by fishing line to a dead _tanifa_. Hoisting the little boy's lavalava on a paddle, they delivered the _tanifa_ to the chiefs of Falealupo. The _tanifa_, Lamositele reports, was large enough to feed three villages.

Bill Rainey began to question Lamositele and several other villagers. Is the _tanifa_ warm-blooded, and does it nurse its young, like a whale? ("No.") Does it have cartilage rather than bone, like a shark? ("No.") Does it have bones like a barracuda, then? ("Yes.") Is it found only in the deep ocean, or does it occur inside the reef? ("The adults are found only in the deep ocean, but the young feed inside the reef.") Can you show us one of the young? ("They are rare, but we will try to catch one for you.") Bill left a carefully labeled plastic bottle filled with preservative with Lamositele in case a baby _tanifa_ was ever captured.

Together with Bill, I raised with the fishermen the issue of poisonous reef fish. At certain seasons in the South Pacific, some species of reef fish feed on toxic plankton and can cause serious neurological disorders and even death in people who eat them. The fishermen told us that such poisoning is rare in Samoa. Later, healer Lemau Seumanutafa pointed out a beach plant, _Scaevola taccada_, that she uses to counteract the effects of poisonous fish.

Bill and Dixie and Thomas and Eva and their children all lived together in our Samoan _fale_. Everyone seemed comfortable, but I noticed that Bill and Dixie cheated a little—they supplemented the woven mats on the hard concrete floor with inflatable backpacking mattresses. Dixie, in particular, seemed to be developing a close relationship with Pela, while Thomas was drawing very close to Lamositele. Each day as we came home from our studies in the forest, I would often see Dixie and Pela sitting together, or Lamositele and Thomas chatting near one of the canoes. Although the condition of

Pela's husband Lilo, who seemed to suffer from Alzheimer's disease, was worsening, I felt increasingly drawn to him. Often he would call out my name—"Koki, Koki"—and I would go and sit by him on the mat. He seemed generally oblivious to the outside world, but was for some reason comforted by my presence. He still was strong enough to walk, with assistance, to the outhouse, but he was becoming increasingly housebound.

Thomas had arrived on the island without knowing the language, but he spent a great deal of time with Lamositele practicing his Samoan while Lamositele practiced his English. Thomas had brought with him from Sweden a small accordion. One evening, after dinner, we gathered all of our Samoan family into the *fale* to hear him play the sweet melancholic music of eastern Europe. The Samoans loved it, and after that evening the villagers frequently asked him to play. Word spread throughout the village, and soon Thomas was in popular demand.

The flow of life in Falealupo had become seductive. Our days began to fit into a steady routine: morning breakfast, followed by research on *Pteropus samoensis* in the rain forest. Evening meal, followed by nocturnal studies of *Pteropus tonganus*. Everything around us was gentle: the people, the flying foxes, the forest—the Samoan ecological triangle. The Samoans protect the forest and revere the flying foxes. Their careful use of the forest guarantees its perpetuation. The flying foxes soar above the rain forest canopy, pollinating the flowers. Later, when the fruits develop, the flying foxes reseed the occasional gaps in the forest. Much like the Samoan healers I study with, the flying foxes serve as physicians to the forest, healing its wounds when a tree falls. The forest provisions the flying foxes with nectar and fruit, and supplies the Samoan people with raw materials for medicine, shelter, canoes, implements, and some of their food.

One evening as we walked along the white sand beach, Thomas commented, "This is really heaven, isn't it?"

I was surprised by the allusion since Thomas is not overtly religious.

"What do you mean, Thomas?"

"I mean Falealupo is really heaven. Everyone is happy, the village is peaceful and beautiful, the people are fantastic, and they have a deeply caring culture. If all the world was like this, we would have heaven on earth."

But unbeknownst at the time to Thomas, there is a dark side to life in Samoa. The next day the village was abuzz with reports of trouble in Salelologa. A young man was drinking beer on the Salelologa wharf and being a bit loud. When an elderly man told him to be quiet, he threw his beer bottle. Though the resultant head wound was superficial, the status of the injured man was not: he was a high chief in Salelologa village. A crowd of thirty to forty villagers chased the young man, who leapt into a pickup truck and sped away. The crowd commandeered a bus and drove off in hot pursuit. A mile down the road, the pickup truck was forced off the road. The mob pulled the driver from his vehicle and beat him to death on the spot. But only after the murder was it discovered that the angry villagers had pursued the wrong vehicle, driven by a man who was unaware of any trouble. The real culprit escaped.

Violence in Samoa can indeed become deadly. The large frames of most villagers are honed into superb physical condition by the rigors of Samoan life. Every day, save Sunday, the village youths typically walk up to ten miles into the mountains to tend their plantations, usually returning with a hundred pounds of taro or bananas balanced on a stick over their shoulders. Samoan fishermen routinely swim through pounding waves to cross the reef, and then dive over forty feet deep to harvest giant *Tridacna* clams. And often after a hard day on the plantation, everyone joins in an impromptu cricket or rugby match. Hand-eye coordination among Samoans is superb: most Samoan youths can kill a chicken at a hundred paces with a stone. In the Apia high court, a stone is regarded as a lethal weapon in the hands of a Samoan. Inhabited by a population large in stature with a vigorous way of life, clean air, and a diet rich in fresh fish, fruits, and vegetables, Samoa has become a major destination for professional sports recruiters throughout the world.

Even given the native strength and dexterity, violence in Samoa would not be so terrible if it existed solely as individual acts— even a large man can be quickly subdued and neutralized by other Samoans. What makes violence so dangerous in the Samoan context is the specter of vendetta. Since familial and village relationships are so broad in Samoa, it is nearly impossible to pick a fight with a single individual. As demonstrated by the tragedy at Salelologa, even an elderly man on an isolated wharf is likely to be surrounded by a multitude of kinsmen and fellow villagers.

Derek Freeman has made much of hostility and aggression in Samoan society, and he has often attacked Margaret Mead's characterization of Samoans as gentle and easygoing. Yet Mead's view was closer to reality: Samoans are almost always gentle and easygoing, and certainly she was right about village youth enjoying an easier passage through adolescence than their American counterparts. Samoans merely employ a different mode of aggression than is typical in most Western industrialized societies. While it can be horrifying, it is nonetheless extremely rare, and bears within it the seeds of future peace. The minimum reducible unit of society in Samoa is not the individual, but the extended family and the village. While conflicts that begin between two individuals can thus immediately broaden into clashes between extended families and entire villages, it is precisely that potential for escalation that can lead to mediation.

Later that night, Lamositele explained to me the Samoan mechanism for reconciliation called *ifoga*. *Ifoga* literally translates as "a lowering," and refers to the abject contrition of an entire village. Performance of an *ifoga* is considered by Samoan courts as partial satisfaction of judicial penalties. Yet the possibility of having an *ifoga* performed on one's behalf serves to deter individuals from engaging in bad behavior, because such actions put one's family, neighbors, and fellow villagers at significant risk.

Before I came to Falealupo, Lamositele recounted, a vehicle driven by a Falealupo man had killed a small boy in Upolu. Word of the tragedy rapidly filtered back to Falealupo, where the village immediately hired a bus to transport the chiefs and village leaders to Upolu. The next morning, as first light began to erode the darkness of an island night, an extraordinary sight was gradually revealed in front of the dead boy's house: sixty Falealupo villagers sat silent and motionless on the road. A fine mat was draped over each of the indistinguishable figures. The significance of the scene was clear to the village: an *ifoga* was in progress, and Falealupo had assumed collective responsibility for the boy's death. There, on the road, in the attitude of absolute humility, sat the entire chiefs' council of Falealupo. And somewhere among them sat the driver responsible for the death.

The father of the dead boy was enraged by the sight. Only a few hours before, he had cradled the cold and broken body of his son. Somewhere in front of him now sat the culprit. He grabbed his machete and headed out of the house to kill the errant driver. But which of the sixty mat-draped figures would he attack first? Fifty-nine of them had no personal involvement in the incident. Which one was the driver?

His vengeful purpose clouded, he was easily restrained by his own village leaders, who urged him to calm down. His own personal loss, they explained, had now been assumed by the entire village. This was no longer a conflict between him and the driver; it was now a conflict between the two villages. Each of his fellow villagers was as aggrieved as if he had lost his own son. He must now allow the entire village to determine issues of vengeance and reconciliation. He could no longer act alone in this matter.

The sun rose higher in the sky and the day became hotter. Still the Falealupo men sat in front of the house. They would not leave— they would bake in the sun until their communal show of remorse was accepted. The village leaders began to reason with the father. "Look, we can't allow their whole village to sit in the sun all day. Yes, it is a terrible tragedy, but Falealupo is still a good village. Some of their chiefs' titles outrank our own. We must forgive. We must find ways to heal this wound."

The man leapt up and reached for his gun. But his attempt was half-hearted, and he fell weeping into the arms of a village orator. "Courage," they urged him. "Our entire village is on trial here. Everything we do here today will be known and reported throughout Samoa. We must try to find it in our hearts to forgive them. We must accept their _ifoga_." Finally, after a few hours, a village chief approached the men on the road.

"Chiefs and orators of Falealupo," he began sonorously, "please enter and accept the hospitality of this sorrowful family and of our grieving village."

Slowly each of the villagers arose from the road, removed his fine mat, shook hands with the village leaders, and embraced the grieving father. The Falealupo orators spoke eloquently of their sorrow at the incident. They begged forgiveness as if they had committed the deed themselves. Their words were backed up with gifts of pigs, money, and the fine mats they wore. Falealupo was forgiven for the offense. The _ifoga_ was accepted.

I asked Lamositele if he wasn't frightened as he sat on the road with his head covered by the mat. What if the father had actually started shooting?

"It is nerve-wracking," Lamositele admitted. "You are never quite sure what is going to happen. That's why each of us took little precautions."

"Like what?" I asked.

"When I am at an _ifoga_, I make a little peephole in the mat covering my head. I was watching really carefully through the peephole when that guy went for his gun."

"Isn't it terrifying to know that you might be required to sacrifice your money, your fine mats, and even place your life on the line because of some offense you didn't commit?" I asked.

"At any moment the call could come through the village to get ready for an _ifoga_. We might not at first even know what it was about. But all of our goods would be immediately surrendered to the village. In an extreme case we might put firewood and *umu* stones in front of us at an *ifoga,* saying in essence that we are presenting our very bodies to be burned. That is the Samoan way."

"Do you have to do an *ifoga* for any villager? What if the offender didn't live here any more? Would the village still have to stand good for their loss?"

"We would perform an *ifoga* for anyone connected to Falealupo, for we bear responsibility for their actions. We certainly would do an *ifoga* for you."

His last statement gave me pause. The thought that an entire village might suddenly be impoverished or subjected to the humiliation of an *ifoga* because of something I might do gave me, for the first time, a glimpse into the unusual web of shared responsibility that Samoans take for granted as part of their daily lives. It was stunning that somehow I had inadvertently become part of that web, and that an entire village would vouch for my good behavior. Searching for a Western analogy, I thought of the names at Lloyd's of London, each of whom pledged their personal fortunes to underwrite insurance policies. Usually the relationship with Lloyd's is profitable, but occasionally, the chips are called in. So it is in Samoa: one's possessions and reputation are constantly placed at risk to guarantee the good behavior of others.

I asked Lamositele one final question: "What if the *ifoga* had not been not accepted? What if that man had actually shot one of you?"

Lamositele paused. "I have never heard of an *ifoga* being rejected. You should ask the old people—perhaps they know of one. But if an *ifoga* were rejected, there would be only one course of action."

"What is that?" I asked.

"War."

The size of our research team enabled us to do some extensive studies of the nocturnal habits of the white-necked flying fox, *Pteropus tonganus.* Not far behind our *fale* were several kapok trees. We knew from Herbert Baker's studies in Africa and Central America that kapok, which has fragrant night-opening flowers, is a favorite plant of bats, which feed on its sugar-rich nectar. Kapok flowers also attract a variety of other organisms, including birds, sphingid moths, and bees; in fact, kapok trees tend to attract a wide range of pollinators wherever they occur. So we thought that we might derive a rough index of the relative diversity of pollinators in Samoa by studying the pollination biology of the kapok tree.

In the afternoon, Dixie started assembling the aluminum sectional poles for the mist nets: fine, long-threaded lattices designed originally to catch birds. I climbed up into a kapok tree and, using duct tape, attached remote-controlled flash units to photograph the bats. Bill snipped some pieces of kapok branches and put them in buckets of water so that we could accurately measure nectar production and time of flower opening. He got the micropipettes and chromatography paper ready for the nectar samples. Thomas prepared to record our data on a laptop computer.

We put fresh batteries in our two night vision devices, a tripod-mounted night scope with a telescopic lens and a pair of night goggles. Originally designed for the military, night vision devices can operate even under faint starlight, but beneath the dense canopy of a tropical rain forest, accessory illumination is required. Since even a faint glimmer of light might alarm bats and other nocturnal animals, we used headlamps fitted with infrared filters and a small battery-powered infrared laser designed especially for us by laser specialist Dr. Paul Farnsworth.

Such precautions made us virtually invisible to humans as well. One evening on a cloudy night, I was sitting quietly to the side of a dark road, watching a banana flower through the night vision device to see if a flying fox would visit it. After an hour or two of fruitless but silent observations, I was preparing to leave, when I saw through the night vision device a rather large Samoan turning the corner, walking entirely by memory. Although I could see him clearly, he could not see me. I wondered if I should alert him to my presence, or if that would only startle him. As the man approached, I turned the infrared laser off, fearing it might damage his eyes with its invisible beam. I sat silently, listening to his footsteps as they drew closer and closer until they were nearly upon me. After he passed, I quietly switched on the laser. It was only then that I saw that in his right hand he carried a machete.

But the people of Falealupo soon became used to having *palagi* scientists wandering around the forest at night with esoteric research equipment. We began our observations of the kapok tree in the afternoon, then worked in shifts, so that we would have a continuous record of all animal visitors to the flowers over a twenty-four-hour period. The branches of the kapok tree come straight off the erect trunk, giving the entire structure a pagoda-like appearance. At 5:00 P.M. the first visitors arrived—a few birds. These proved to be not pollinators, however, but floral robbers, piercing the unopened flowers at the base and drinking the sweet nectar. The flowers themselves did not open until 7:30 P.M., when they began to exude a fragrant aroma. Shortly before the flowers opened, however, the first white-necked flying foxes had begun to arrive. These were not the Samoan flying foxes, *Pteropus samoensis,* which we never did record visiting a kapok tree, but the colonial, night-foraging *Pteropus tonganus.* The flying foxes flew to separate branches and waited for the flowers to open. With their jet black bodies and white collars, they looked like young suitors in tuxedos waiting expectantly for their prom dates. All was tranquil until some late arrivals sparked noisy squabbling. Two flying foxes on the same branch do not tranquility make, and as soon as the second one arrived, there commenced acrimonious bickering, followed by a kind of boxing, in which opponents tried to jab each other with their long, extended thumbs. These contests ended with the abrupt departure of one or the other of the animals: usually the latecomer lost. Apparently flying foxes, like many humans, find existence untenable without their fellows, yet seem incapable of living together in harmony.

Given this lack of flying fox amity in nature, we were astonished at the docility of captured animals. When Dixie retrieved a white-necked flying fox from the mist net, she put on gloves in the expectation that the animal would be frightened and aggressive. But the bat was gentle, patiently allowing Dixie to untangle the net from its wings and feet. After freeing it, Dixie hung the bat on her shoulder, where it remained without restraint, looking up at us with its great brown eyes. Clearly it had descended from generations of island creatures that knew no natural enemy. No hawk or cat, eagle or snake had ever threatened its ancestors; nor had its ancestors ever threatened any other living thing. Despite their occasional noisy quarrels, flying foxes are vegetarians, their diet consisting solely of fruit and nectar. There was thus nothing in the flying fox's inherited behavioral repertoire to suggest that Dixie might represent a danger.

The animal seemed fascinated with our activities. Flying foxes do not echolocate as other bats do, and therefore must of necessity observe VFR—visual flight rules. Nature has fortunately equipped them with the biological equivalent of a night vision device: large eyes that see well on dark, starlit nights. But perhaps most striking are their large, dilated pupils, equaled in size only by those of the lemurs of Madagascar. To have such a gentle but wild creature scrutinizing you under no constraint other than its own curiosity is to feel the weight of God bearing upon your soul. That night, as I worked under the watchful eye of the little flying fox, I felt my mother's brown eyes caressing me again. The only experience I can compare to that evening is the afternoon we swam as a family with wild dolphins in a remote part of Western Australia. Our children went to pet two dolphins that approached the beach, but to their disappointment, the pair soon disappeared. Five minutes later, though, the dolphins returned, this time with their newborn baby. I wonder if we could learn sufficient humility swimming with dolphins or gazing into the eyes of flying foxes, if we might then discover unimagined beauty.

After posing for several photographs, and even submitting without protest to having her blood drawn by Bill, the flying fox, her dignity still intact, flapped off into the cool tropical night. When our vigil came to an end, the dawn revealed a research site littered with cameras, tripods, and laptop computers, on top of which had descended, like a late spring snow, the spent white corollas of the kapok flowers. The data we collected that night caused us to wonder if island rain forests differ in their pollination biology from continental rain forests. Only a single animal species, *Pteropus tonganus,* pollinated kapok in Samoa. Not only did this study confirm our suspicion of the presence of few species of pollinators in Samoa, but it was also consistent with our belief that *Pteropus samoensis* forages principally in the daytime—careful observations made with night vision devices failed to reveal any visits to kapok trees by that species.

But negative evidence holds little water in scientific circles. Though we did not see Samoan flying foxes visiting kapok flowers at night, we could not claim that they didn't do it under other circumstances, or did not forage at night on different plants. To corroborate the diurnal foraging pattern of Samoan flying foxes, it was necessary to watch a breeding pair very carefully over a twenty-four-hour period. We decided to hike to the *Pteropus samoensis* roost on Mount Fuionō, where Thomas and I would make the necessary observations in the daytime, and Bill and Dixie would take the night shift.

Our plans alarmed Pela and her family. Since Lilo had grown increasingly infirm, Pela had become the anchor of the family, and few major decisions were made without her approval. Samoans are reluctant to walk at night, when *aitu,* or spirits, might be afoot, and she did not want Bill and Dixie spending the night alone in the forest.

"Do they have to go up Mount Fuionō at night?" Pela asked. "Can't they be content with staying up there in the daytime?"

"They need to stay up there for twenty-four hours to make certain that Samoan flying foxes aren't active at night."

"But we know that already. Why waste time proving the obvious? Furthermore, how will they be able to see anything up there? There is only a quarter moon tonight."

"They will take our night vision devices, Pela, that allow us to see in conditions of near total darkness. They do this sort of thing all the time. All of us do. You have to remember that in our culture, we are not afraid of the forest at night."

"Yes, but in America, you don't have *aitu*. Let me send a few of the boys with them, just to make sure they will be O.K."

"No, Pela, they'll be fine. I'm going to hike up there at dawn, and they'll relieve me late in the afternoon and stay through the night."

Pela continued to fret about the planned expedition as we carefully packed all of the necessary equipment, including cameras, headlamps, mosquito repellent, night vision devices, notebooks, and two foam pads. Our plan was to maintain a twenty-four-hour vigil above the *Pteropus samoensis* roost, recording the bats' activities and foraging patterns. We were particularly interested to see whether they stayed asleep during the night or sometimes flew about. Clearly their eye morphology argued for strictly diurnal behavior, as their pupil diameter, unlike that of *Pteropus tonganus*, is almost eaglelike in its narrowness. Hiking the steep slope, I found a flat place on the summit, and settled in above the roost.

The Samoan flying fox, *Pteropus samoensis,* has a demeanor very different from that of *Pteropus tonganus.* Unlike the gentle, docile white-necked flying fox, the former is haughty and condescending. One would never wish to hang a *Pteropus samoensis* on one's shoulder. I once captured an adult in the field, and it tried to bite my finger off. Like a proud prisoner of conscience, it would refuse to feed in its cage whenever I came into sight.

When I arrived on top of Mount Fuionō to relieve Thomas, the male *Pteropus samoensis* initiated a territorial display by hanging by one foot while carelessly extending his lower wing below his body. Roughly interpreted, this meant, "O.K., human, I know you are here, but don't forget who owns this place."

I carefully recorded in my notebook the flights of the flying foxes. Unlike any other bat in the world, *Pteropus samoensis* appears to fly for the sheer joy of it. Leaving their baby behind in the roost tree, the male and female soared on the thermals high above the forest canopy, playing games of follow-the-leader. In between their frolics, they landed on trees to feed on fruit. They flew beneath the forest canopy and disappeared from sight, only to soon reappear, soaring high into the air. Time passed quickly, and it didn't seem long until the sounds of Bill and Dixie climbing the hill informed me that my shift was up. They arrived cheerily, we chatted a bit, and I set off alone down the hill.

That night at dinner in our *fale*, Pela seemed upbeat, and her worry over Bill and Dixie spending the night alone in the forest seemed to have abated. It wasn't until later that I learned the probable source of her solace.

As the night grew late on top of Mount Fuionō, Bill and Dixie, watching carefully through night vision devices, saw the pair of Samoan flying foxes wrap themselves up in their wings and go to sleep. After several hours of watching the bats sleeping, Bill and Dixie decided that together they could take a short break from their vigil, as nothing looked like it was going to change. Their little perch on top of Mount Fuionō, primitive as it was, offered them the first privacy they had experienced since arriving in Samoa two weeks earlier, and they took full advantage of it.

The resultant lapse of twenty or thirty minutes in the logbook went unnoticed by me, but Bill and Dixie became the subject of much good-natured teasing by the villagers the next day. As they soon discovered, privacy is usually illusory in Samoa, and I still wonder if Pela sent someone up the mountain that night to secretly keep an eye on them. If so, according to Dixie, they must have gotten an eyeful.

But the information Bill and Dixie brought back was important. As we suspected, the *Pteropus samoensis* pair slept throughout the night, resuming their activity only the next morning. The Samoan flying fox seemed as unique in its diurnal behavior as we had claimed. But their proclivity to forage during the day, coupled with their innate curiosity, makes these flying foxes particularly vulnerable to poaching. And their low reproductive rate and rather exacting habitat requirements prevent the species from rapidly recovering from population destruction.

"Paul, I'm afraid you were right," Dixie said one day. "Unless the Fish and Wildlife Service bans their import into Guam, there will be little chance of survival for the Samoan flying fox. But given the political realities of the Service, I can't foresee them taking any action to list it as an endangered species."

"What about international intervention?" I asked. "After all, the importation of flying foxes from around the Pacific into Guam constitutes international traffic in endangered species. Shipping crocodile skins and endangered bird feathers across international borders is prohibited. Isn't there something that the international community can do to control commerce in flying foxes?"

"Indeed there is," Bill replied. "The Convention on International Trade in Endangered Species (CITES) is a coalition of over a hundred countries that regulates such traffic."

"So why don't we get them to act? Perhaps they can help."

"The problem," Bill said, "is that the United States is represented at CITES by your friends at the Fish and Wildlife Service. And although there are still many good people in the Service, I don't think that they could go against the recommendations of the Honolulu office in this matter."

How could we expect the U.S. government to request international proscriptions of trade in the species when its own Fish and Wildlife Service refused to take any protective action domestically? If we failed in our attempt to create a national park in American Samoa, our only chance to save the flying foxes would be for another signatory nation to CITES to take up the cause. Thomas smiled.

The next day, while Bill and Dixie busied themselves with preparations for some flying fox observations, Thomas and I walked the path toward Fagalele beach. The trail wandered around the twisted roots of a large *ifi* tree and through a small clump of white forest ginger. Samoans use the copious, fragrant sap from the red floral heads as a shampoo and hair conditioner. Near the beach, we saw a stand of *Pandanus* trees with their mysterious stilt roots. We collected a few plants, but did not see any flying foxes or birds. We walked back toward the village. At the edge of the forest, some village youths pointed out to us a small tree with round yellow fruits—*Diospyros samoensis*—that they use for fun to adorn their skin, like tattoo decals for Western children. I took sufficient material to make a pharmacological collection for the National Cancer Institute.

As Thomas and I continued back toward the village, I thought about how lucky I was to be in Falealupo, a true refuge from trouble. There were commercial bat killers elsewhere in Samoa, but they hadn't come to Falealupo. Despite the erosion of traditional Samoan culture in Pago Pago and Apia, the chiefs continued to govern Falealupo in accordance with ancestral wisdom accumulated over centuries. The rain forest in Samoa as a whole was disappearing, and I knew that the Falealupo forest was threatened because of the money needed for the new school. But the rain forest seemed such an integral part of the village, and of my life, that emotionally I could scarcely imagine that it could ever disappear. Somehow, the village would find a way to protect the forest. Whatever happened to the rest of Samoa, Falealupo would remain my home forever. I had achieved *agatonu*—correspondence—with the village.

Thomas and I walked into our *fale*. The mood was oddly tense. Lamositele looked up at me and, in a voice filled with apprehension, said, "The loggers have arrived. They're cutting the rain forest."

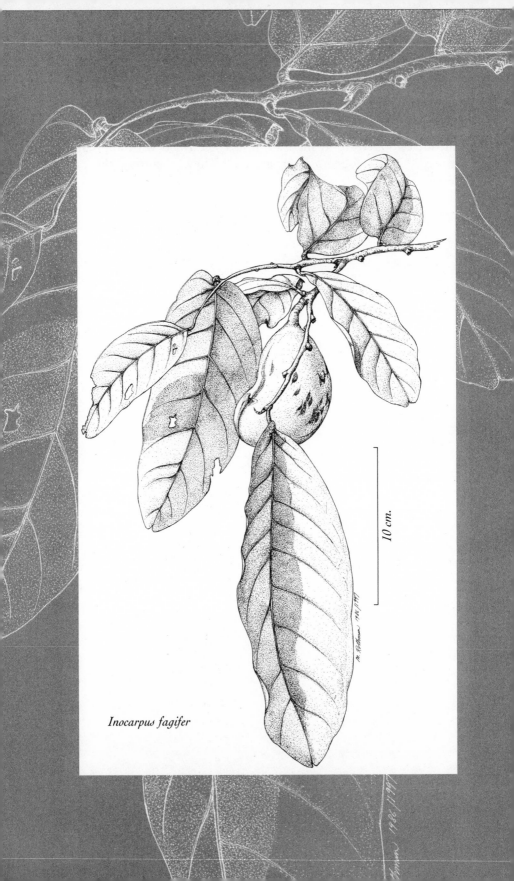

Inocarpus fagifer

CHAPTER 7

Crisis

An immense sickness flooded
over me suddenly. . . . The thing
which was waiting on the alert,
it has pounced on me, it flows
through me, I am filled with it.

Jean-Paul Sartre, *Nausea*

L ilo Manuele climbed into the front seat of the van while Thomas, Bill, Dixie, and several other villagers jumped into the back. The wheels spun briefly in the white sand in front of our *fale* until I remembered to engage the four-wheel drive. With an abrupt jump, the vehicle started down the road. Lilo was speaking rapidly to me in Samoan:

"Koki, the people who have seen it say it is really horrific . . ."

Dixie interrupted in English, "Why are you dragging us up there, Paul? We're right in the middle of analyzing the nectar samples. Don't you think you and Thomas are sufficient?"

I answered Dixie first. "I know this is inconvenient, but I've got a feeling that we all need to go up there, to see what is happening. We'll get you back to the research just as soon as we can."

I then spoke quietly in Samoan to Lilo Manuele. "Why has the logging started now? I thought the village was trying to delay it."

"They were," Lilo began, "but the government gave the village an ultimatum. Either we build the new school within nine months, or they will shut our existing school down and withdraw all the teachers. So the village borrowed money from the logging company to start construction in exchange for logging rights, and I guess the loggers decided this week to extract their debt."

We drove up the bumpy road in silence, each of us absorbed in thought. At the road below Mount Fuionō, the volcanic cinder cone with the Samoan flying fox roost, I paused to look for the flying foxes. Usually the pair could be seen circling the little hill at this time of the afternoon. But we couldn't see them anywhere.

We continued up the road, entering the primary rain forest. Again, all looked normal. Large, cablelike lianas hung from the tall canopy-level trees. The morning light, filtered through gaps in the canopy, illuminated a few odd bird-nest ferns growing on the ground. But the forest was strangely quiet—no birds were singing—and there seemed to be a mist or dust in the air that caused the morning light to appear like piercing daggers in the darkened forest.

"Doesn't look like they've begun logging here," I said to Lilo.

"No, it's farther up, and to the right of the main road."

We continued through the lowland rain forest, perhaps the largest of its kind in Samoa. Two children carrying bananas on their backs were the only humans we saw. But just before we exited the forest, a large logging truck came roaring up behind us, forcing me off to the side of the road.

"I guess we passed it," Lilo said sheepishly.

I drove out of the forest and turned the van around in the little dirt lane that led to a nearby *fale*. We went back about four hundred meters and turned onto a dirt track that led to a little *fale* in the forest. Soon we could hear the sound of heavy equipment.

The transition from rain forest to clear-cut was as abrupt as the transition from heaven to hell. Our ears were assaulted by the noise of the chain saws, diesel loaders, bulldozers, and the roar of the engines of the huge diesel trucks. The earth itself, the very dirt under our van, was raw and wounded, bleeding from deep ruts and gouges. Once stately forest giants were being heaped into piles by the bulldozer blades. The logging company was not content to merely cut individual trees: at the edge of the clearing, a bulldozer pushed the border of the forest back like a curtain. From behind us, the sudden crash of a huge tree, lianas and vines trailing from it, diverted our attention. To the right, a large loader, looking like a giant insect with mandibles in attack position, grasped the logs and dropped them onto the beds of the diesel trucks. Once loaded, each truck raced down the road as if fleeing the scene of a hit-and-run accident.

A Samoan foreman, seeing us, walked over to ask if we needed assistance. I told him we were just watching.

Watching, indeed, for the shock was so great that I initially felt a stunned detachment, as if I were viewing the autopsy of a member of my own family. Forests are cut every day, and I tried to pretend that nothing unusual was happening here. The technique of logging was not novel: bulldozers pushing down all the smaller trees and scrub so that men with chain saws could more easily cut down the few desirable trees. The equipment was not unique: the surgical mask, the needles, the forceps, the smell of formalin, a bit of blood on the sheet—the bulldozers, the chain saws, the noise of the huge trucks, nothing was new here, all comfortably familiar. I tried to watch, but the scene became blurry. I looked over at Dixie. She was weeping too.

"Paul, this is awful. You've got to get us out of here."

Bill's voice cracked when he spoke. "I can't believe this."

Lilo had tears in his eyes. "It's like an atomic bomb. This is terrible. The forest will never recover, and the ground will be useless forever."

Dixie spoke with greater urgency. "Paul, I can't bear this any longer," she implored. "You've got to . . . "

Her words were cut short by the crash of another tree just in front of us.

I insisted that we stay to witness the destruction. I wanted this scene burned into each of our minds: the death of the last lowland rain forest in Samoa. Not only had the trees been cut, but the very soil was being defiled, scooped out of the way. Every way I turned, destruction: broken and shattered trees on the ground, piles of soil and debris cloaked in acrid diesel fumes. I could see some Samoan children standing far off at the edge of the forest, as trucks laden with massive trees, their heritage, barreled past them on the dirt track.

Devastation. No other word could describe the scene before us. A setting that was once transcendent with gentle splendor was now the abode of evil. I looked at the retreating edge of the forest, beyond the bulldozers, toward the ferns, the lianas and vines hanging from the banyan trees. This forest was completely vulnerable to the onslaught of the loggers; it could only wait silently for the next phase of destruction to begin.

"Paul," Dixie exclaimed, "that bulldozer is heading in the direction of the flying fox roost!"

One of the bulldozers was beginning to scoop out a swath in the forest in the direction of Mount Fuionō, three thousand meters away. All three sides of the Samoan ecological triangle were now endangered, I realized: the flying foxes in the roost, the forest that was disappearing, and the Samoan culture shared by Lilo and the village children. And as yet another truck filled with logs drove away, I felt completely helpless. In one day of activity, the logging company had already destroyed one hundred acres. At that rate, the forest would be completely gone in three hundred days—less than a year.

As I witnessed the destruction of the forest, I felt as if I were again watching my mother die. I could do nothing then—the best doctors had been employed, but to no avail. God had been importuned with fasting and prayer, but to no avail. Comfort and care and love had been lavished, but to no avail. I remembered so clearly the moment when I received the results of the last blood tests, the cancer cell counts that incontrovertibly proved that no agent, medicinal or divine, would save my mother from a cruel and agonizing death. I sat stunned on the hill above my home, then drove down to the town to talk with the bank president, and subsequently a loan officer. I left with a check, and arranged to have a hot tub installed immediately behind my mother's home. "Even if it gives her a few moments of relief from the pain, it will be worth it," I told Barbara. "And besides, I have to do something."

So I stood now, completely helpless, watching the loggers rip the heart out of the Samoan rain forest. Never before, except at the death of my mother, had I felt so small, helpless, insignificant, inconsequential. Once again I felt awash in despair.

Yet another truck, laden with timber, moved jerkily down the road toward the sawmill. I thought of the final drive with my mother to the hospital. Before departing her home, I called the oncologist, who cautioned me not to hurry on the fifty-mile drive to the hospital. "Paul," he said gently, "there's little we can do for her."

I lifted her frail body from her bed and cradled her in my arms to the front seat of my old Chevy station wagon. The red vinyl seat was cold in the November air, so I pulled a woven orange cap I had in the car over her ears. Dad climbed into the back seat. I gently accelerated, knowing that even the slightest change in velocity shot pains like lightning through her emaciated body. My mother, the woman who had given me life and love, who had patiently placed my tiny hands on her microscope, who had untangled my fishing lines, who had always smiled so wisely at me, clung with whitened knuckles to the car seat. An hour later we finally arrived at L.D.S. Hospital in Salt Lake City, where we were met by nurses with a wheelchair.

"That's the longest trip I ever made in my life," my mother said in a hoarse whisper. She then lapsed into a coma.

As I watched the Falealupo rain forest die, I fumbled with my camera to photograph the destruction, but I could hardly see through the viewfinder to focus the lens. What could I do? The logging had already started, moving at such a pace that the entire thirty-thousand-acre forest would be completely destroyed in less than a year. Another logging truck screamed past us as we walked back toward the van. I was completely shattered for the second time in my life.

As I watched the chain saws chew their way through the Falealupo forest, I felt deep inside that the fate of all the forests of Samoa, and perhaps other islands of the tropical Pacific, would be decided at this place and at this time. If the Falealupo villagers, resident on their peninsula for centuries, could not protect their forest, then no rain forest in all of Polynesia was safe for long. Yet any battle to save the forest seemed doomed before it began. This was no academic issue to be debated in a weekly seminar, or discussed at a symposium with economists and resource specialists. There was no time to try to enlist overseas environmentalists to save a place of which few had heard before. There was little history of conservation in Samoa, and the rain forest was dying—now. The flying foxes would inevitably follow, as would, ultimately, the Samoan culture. This was no time for academic considerations. My head was reeling, but I knew I must follow my heart. Silently I made my decision. At Falealupo, I would make my stand: I would do everything in my power to stop this logging.

As we drove back to the village, I spoke quietly in Samoan to Lilo. "I must meet with the village chiefs. We have to do something to stop this logging, and I've got an idea. But we've got to move fast: the loggers will be at the flying fox roost soon. How soon do you think I could have an opportunity to meet with the chiefs?"

"Have you heard about the investiture ceremony on Saturday?"

I had, because Lamositele's family was involved, but I knew little about the sequence of events, nearly all of which were restricted to chiefs.

"All of the chiefs will gather," Lilo explained. "Two men will receive the same chiefly title: Foa'imea. Come at the beginning of the ceremony, and then you can speak to the entire chiefs' council."

The next morning, the loggers were working again. By the end of the day, another hundred acres would be gone. I drove to Asau on an urgent mission.

The branch post office in Asau had a telephone, which sometimes worked. Overseas calls from Asau were not common, but once I had driven Michael up there to call his wife in Manhattan. After a short wait, he had actually gotten through. "Please let that phone work for me today," I silently prayed.

The telephone exchange at the Asau post office looked like a scene from a Mad Max movie. Wires were strewn throughout the dingy concrete room, leading to a metal box without a cover. The phone itself was missing its base. Sprawled full-length on a desk, wearing jeans and an unbuttoned orange shirt, was the sleeping form of the telephone operator.

"*Sole!*" I yelled, waking him up. "I need to call the USA." Direct-dialed overseas calls are not possible from Asau, so we had to wait an eternity for the Apia operator to come on line. I waited at the counter outside the building.

"O.K.," the young man finally yelled out to me, "pick the phone up now."

I cradled the bottom of the telephone with my hand so the inner mechanism wouldn't fall out. Through the tiny speaker, I could hear the Apia operator ask me where I wished to call.

"Milwaukee. It's in the USA."

I quickly gave her the telephone number, area code first.

"And whom do you wish to speak to?" she asked.

"Verne Read."

I prayed he was there. I needed Verne's help. And I needed it immediately. But did I have enough of a relationship with Verne to ask for his assistance? A fly lazily buzzed across the cracked Formica counter inside. The telephone operator was sprawled again across his desk, his eyes closed. I was suddenly impatient with Samoa. The logging was going on now! Why did I have to wait forever, holding some cannibalized phone in my hand? I listened carefully to the sounds of the ether. After several long minutes, I heard the voice of the operator in Apia.

"Go ahead, Mr. Cox. Your party is on the line."

"Verne, this is Paul Cox in Samoa. Can you hear me?" I shouted, listening to the satellite echo of my own voice.

"You're coming through loud and clear, Paul. How are you?" Verne replied.

"Verne, I need your help. And I need it badly."

"Yes?" he replied.

"Do you remember how I told you that I was working in Savaii in Falealupo village—the village that had valiantly turned away the sawmill's overtures for all these years?"

"Yes, Paul."

"Well, the loggers have now arrived and they're busy hacking down the forest. It's terrible! Everyone is very upset. They're destroying the largest remaining lowland rain forest in all of Samoa. But there's a chance we can stop it."

"What do you have in mind?"

"Verne, I need you to guarantee a loan. They're cutting down the forest to pay for an elementary school. If they don't build the school immediately, the government won't allow the village children entrance into the country's educational system. So they've licensed their forest to the logging company, with the income going to pay for a loan from the Development Bank of Western Samoa to build their school."

"How much is the loan for?" Verne asked.

"The best I can figure, it's about $65,000 U.S. I plan to offer to personally assume the loan if they can stop the logging immediately. I haven't talked with the village about this, and I don't know whether they'll accept my offer or not. But if they do, I'll need a bit of time to liquidate my assets and raise money. I'll do anything to get it, but I need some time. What I'm asking for, Verne, is six months. Could you guarantee the loan payments for six months, so that I have a window to try to come up with the rest of the money?"

"How much will that cost?" Verne asked.

"I'm not sure what the monthly payments are, but I'm guessing around $600 to $700 per month. If you could pledge $4,000, Barbara and I could make up any shortfall."

There was a brief pause on the line. "Paul—you're covered. Go for it." Verne said. "And good luck."

Thank God for Verne and Marion Read. I drove to the little shack that housed the airline office in Asau, and booked myself on the Monday morning flight to Apia. If the village agreed to have the logging immediately stopped, I needed to be at the Development Bank office then to assume the loan. But deep inside, I was terrified. I knew that getting Verne's help was the easy part. The difficult part would be getting the village to agree to my offer. Samoans treasure their land above all else, and they had been frequently exploited in land deals with foreigners. Asking the villagers to trust me, a foreigner, on a land issue would be extraordinarily difficult.

When I awoke before first light the next morning, the sweet smell of _umu_-baked taro and pig filled the air. The last embers of the fire glowed in the cook hut behind our house. Young men, speaking in low whispers, spread the large, red-hot stones about with tongs made from the midribs of coconut leaves. Ready at the side of the hot rocks were taro, _ta'amu, palusami,_ the chestnutlike kernels of the _ifi_ tree _(Inocarpus fagifer)_, plucked chickens wrapped in bird-nest ferns, and several large fish with baskets woven around them. These were placed on the bed of hot stones, and then more hot stones were placed on top. The entire mound was then covered with fresh banana leaves and old mats, exuding a sweet-smelling smoke throughout the village.

From inside my mosquito net, I could see the dugout canoes pulled up on the white sand beach, and beyond them, a dull streak of orange across the horizon. I knelt in the darkness for morning prayer. But the words did not come, and I struggled in silence. As a young missionary nearly twenty years before, I had, for a period of several months, given up speaking English entirely in my struggle to master the Samoan language. My prayers, filtered through my small Samoan vocabulary, were most difficult. But now, although fluent in both languages, I could scarcely express myself. A Samoan chiefly saying passed through my mind—*"Ua mo'om'o fa'alupe ona o naumati"* ("I thirst for words as a pigeon suffers in the desert")—and with it came utterance:

"Heavenly Father, please save this forest, each plant, each animal, and this village that has protected it. Today of all days of my life, please give me the words to speak to the village. Let me speak the chiefly words with power and emotion. Please touch the hearts of the chiefs to hear my plea . . ."

I arose silently and quietly folded the mosquito net. As I stepped outside onto the white sand, shafts of light filtered through the smoky haze enveloping the village. I could see other families also preparing their *umu*. The morning seemed ripe for either loss or redemption. Either the village would make the leap of faith necessary to trust a *palagi* and accept my proposal, or they would reject it, and the rain forest would be lost forever. I walked slowly to the sea, trembling inside. Theodore Roethke's poem flashed through my mind:

I wake to sleep, and take my waking slow
I feel my fate in what I cannot fear
I learn by going where I have to go . . .

Of those close beside me, which are you?
God bless the Ground! I shall walk softly there,
And learn by going where I have to go . . .

This shaking keeps me steady. I should know
What falls away is always. And is near.
I wake to sleep and take my waking slow.
I learn by going where I have to go.[55]

I approached the sea, and let the waves wash my feet. The inexorable, gentle motion of the water comforted me. Yesterday, as I had watched the rain forest silently approach its death, betrayed for a few pieces of silver, I was disconsolate. But this morning, the sea itself gave me courage. As I stood in the shallows, my composure seemed to grow, and my mind began to fill with the chiefly rhetoric Aumalosi had taught me so many years ago.

After mentally rehearsing for ten or fifteen minutes what I would say, I returned to our *fale*, where I found Lamositele working.

"Lamositele, do you have any suggestions for me today as I approach the village? Do you think I have a chance to sway them?" I asked.

Lamositele's voice became unusually solemn. "It will be very hard. We know and trust you, but land issues are difficult, the hardest ones we have. Any proposal involving land is bound to arouse suspicion."

"But you and Lilo will be there to help me if I have trouble, won't you?

"Koki, I'm sorry, but we can't go. If we were there, the village might suspect that our family put you up to this, and that we are somehow trying to seize village land. It's important for us to stay completely out of the matter."

I stood looking into Lamositele's deep brown eyes, and knew, as always, that he was telling me the truth: given such an unprecedented offer, the villagers might suspect that I represented a plot by Pela and Lilo's family. But still I wanted him to sit next to me, whispering advice and support, since this would be the first chiefs' council I had ever attended. Only chiefs are allowed in such gatherings, and it was extraordinary that I, as an untitled man, was even given a chance to address them.

Pela's family was scurrying about, moving my research gear out of the hut. "What's going on?" I asked one of the young men.

"We're having a family meeting before the ceremony and need to use your _fale_."

Fa'asaina and other young women vigorously swept the sand out of the _fale_ with long brooms made from coconut leaf midribs before spreading mats to cover the hard cement floor. Soon a battered old red pickup truck, heavily laden with mats, pigs, and people, pulled up in front of the _fale_. Part of Lamositele's extended family had arrived. I sat with Lamositele at the back of the _fale,_ as custom dictates when receiving guests. We recited the entrance rhetoric, and then Lamositele introduced me to his guests, some rather distant relations from Asau.

Some young men started to unload large rolls of mats from the back of the truck. I didn't pay much attention to their activities until the first roll was brought into the _fale_. With a start, I saw that they were not ordinary mats, like the ones we were sitting on, but _i'e toga_, or fine mats. Each fine mat must have taken a woman up to a year of solid work to complete. Looking at the rolls, I estimated that there must be thirty or forty fine mats in total. Almost half of a century of human labor was being laid in the _fale_ in front of me.

"_Sa'ō, falā lelei!_" ("Wonderful, good mats!") Lamositele cried in appreciation. Soon Fa'asaina and the other women behind the house took up the cry: "_Sa'ō, falā lelei!_"

I was touched. Lamositele's distant relatives had brought the mats to assist his family in preparing for the investiture ceremony.

Since Lamositele was distantly related to one of the young men who would be granted a chief's title that day, he was obligated to assemble fine mats, pigs, cases of canned goods, and money to aid in hosting visitors to the ceremony. Just as in Western weddings, where the family of the newlyweds provides an often luxurious dinner for the invited guests, Samoans provide goods for their guests on all important occasions: weddings, funerals, or chief investitures. But Samoans are not content to merely feed their guests during the celebration. They make sure that there are plenty of mats, money, and pigs to take home, which are distributed by chiefs representing the village in which the event is held. Extravagance on such occasions is the rule. The resultant burden of such ceremonial occasions, or *fa'alavelave,* would be unbearable on a family if their relatives did not pitch in with material assistance.

Lamositele sat cross-legged, counting the fine mats as they were unfolded. *"Sa'ō, falā lelei!"* he said. "Thanks for the fine mats and for the pigs. And now, how about the money?"

I was taken aback by the directness of his request, which seemed to be out of character. His relatives seemed wary. "We have one hundred *tālā* to contribute to the event," one of the men announced, handing Lamositele's son a roll of bills, the rough equivalent of a year's income for a family in Falealupo.

"One hundred *tālā!* One hundred *tālā!"* Lamositele yelled, visibly angered. "One hundred *tālā*—what can I do with that? That's not nearly enough!"

I had never before heard Lamositele raise his voice. Such an accusation as he made, particularly regarding a gift, contravened all I had previously learned about Samoan customs of respect. The relative began to mumble an apology. "Look, it was the best we could do. We've had a rough year, and this is all the money we could raise."

Lamositele cut him off. "One hundred *tālā!* That's insufficient! You know good and well the mats and pigs don't mean anything, unless the money is sufficient! We need more money!"

The relative reached into a weathered brown satchel, and pulled out another roll of bills. "O.K., O.K., here's another one hundred *tālā*. But that is really all we can spare."

"Two hundred *tālā!* Two hundred *tālā!"* Lamositele yelled. "That's not nearly enough! You know the ceremony won't be right, unless the money is right! How can we have a ceremony with this?"

The relative squirmed under Lamositele's rebuke. Uncomfortable witnessing what appeared to be a family squabble, I slipped out the back of the hut. I found Lilo at the young men's *fale.*

"What's going on with Lamositele?" I asked Lilo. "Why is he so angry with your relatives? Why isn't he grateful for all of the mats and pigs and money they've given him?"

"Koki, you don't really think Lamositele is angry, do you?" Lilo asked me.

"But he's screaming at your relatives. Why is he being so rude?"

"Koki—you don't understand. That's just the way it always happens in our custom. Lamositele is acting like a chief. He needs to put on a big show, and yelling like that makes it clear to the neighbors and anyone within earshot that our family takes this event _very_ seriously. It dignifies the event."

"But does he have to berate your relatives?" I asked.

"That's just the way chiefs behave. Lamositele isn't really mad. But when chiefs get together, that's the way they act sometimes. It makes the ceremony more _mamalu_—more dignified."

I had a sinking feeling. How could I have worked in Samoa for such a long period and never have realized this essential part of the custom? Was I really ready to meet with the village council that afternoon? And what if they screamed at me? Was I supposed to scream back?

I was filled with apprehension. Clearly my understanding of Samoan chiefly custom was insufficient, or at best, seriously flawed. Maybe the times the Samoans had told me that I spoke eloquently amounted to little more than flattery. They must have been humoring me, as they do new tourists who learn a few simple greetings. How could I persuade the chiefs' council to save their rain forest at the risk of their children's education? Despite all my private pretensions and arrogance, I had to face the fact that I was not a Samoan. Yes, I did know some chiefly language, perhaps far more than other _palagi_. And yet, I wondered if I didn't come off to the Samoans like a trained parrot, a mere cultural novelty that might have brief amusement value, but could effect no meaningful change.

What I did know about Samoan custom, though, warned me that I might face serious trouble in the hours ahead. The most contentious topic that can be discussed in all of Samoa is land, and that was exactly the topic I had to raise. I had somehow to convince the village to trust me, a foreigner, and stop the logging, which was being carried out with heavy machinery operated by native Samoans.

After leaving Lilo Manuele, I returned to the sea. I stood on the beach, again mentally rehearsing the little speech I had prepared. Would my words be sufficient to sway an entire village?

A young girl came running up to me. "Koki! Lilo says it's time that you go to the chiefs' meeting. They're waiting for you.

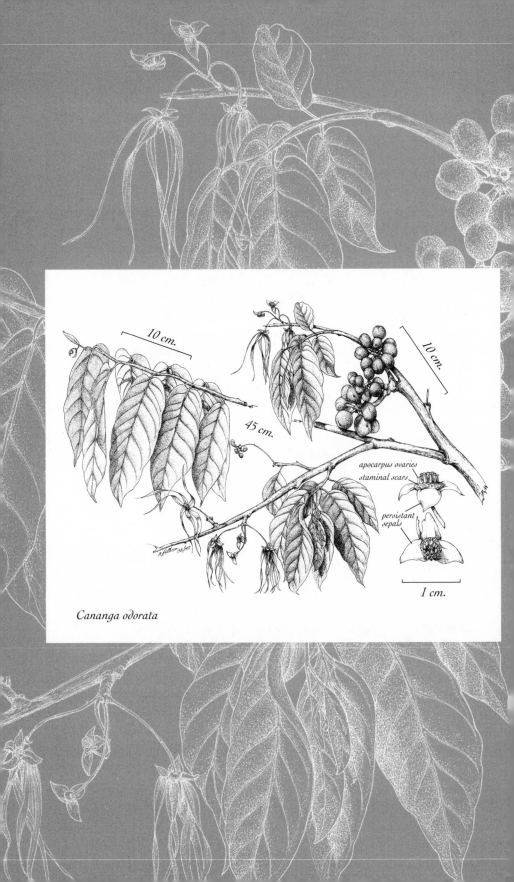

10 cm.

10 cm.

45 cm.

apocarpus ovaries
staminal scars

persistant
sepals

1 cm.

Cananga odorata

CHAPTER 8

Conflict

*Every age, every culture, every custom
and tradition has its own character, its
own weakness and its own strength, its
beauties and ugliness; accepts certain
sufferings as matters of course, puts up
patiently with certain evils. Human life
is reduced to real suffering, to hell,
only when two ages, two cultures and
religions overlap. . . . Now there are
times when a whole generation is caught
in this way between two ages, two
modes of life, with the consequence that
it loses all power to understand itself.*

Herman Hesse,
Steppenwolf

fter carefully adjusting the formal lavalava and red
Pandanus lei that Lamositele had lent me for the occasion, I picked up my long, slender kava root and
walked along the beach to the north side of the village, where a tiny and rustic *fale* on crooked stilts served as the
gathering point for the chiefs. The poles were decorated with red hibiscus flowers and pink torch ginger, while the stilts were adorned
with coconut fronds. I could see about twenty large men packed inside, each stripped to the waist. Among them were both men holding the title of Fuiono: Fuiono Senio, short, handsome, but aggressive in demeanor, and tall, slender Fuiono Mase'ese'e, who seemed
possessed of ineffable grace and dignity. As I looked inside and saw
all of the paramount chiefs of the village there, I had a heightened
sense of being from a different culture and even a different time. Indeed, I was the only man present not to hold chiefly rank. I climbed
the notches in the coconut trunk ladder, stepped inside, and sat
down on the mat. I heard someone murmuring in the back, "What's
Koki doing here? Doesn't he know this is a chiefs' meeting?"

The men's bodies glistened with scented coconut oil. A few wore red *Pandanus* leis like the one Lilo had given me, while Fuiono Mase'ese'e wore one made of tiny red *nonu* fruit, which he occasionally nibbled on. He looked formidable, inscrutable, and very Polynesian, like a statue from Easter Island. He began speaking in the loud, clear voice characteristic of high orators:

Ia, ia sūsū maia lau sūsūga	Highest welcome, Koki,
Koki, o le ali'i saienisi mai le	the scientist from the great
atunu'u tele o Amerika!	country of America!

I responded by reciting the village *fa'alupega,* the list of paramount chiefs of Falealupo village:

Ia, ia afio lava le pa'ia	*Highest greetings to the sanctity*
maualuga o le maota nei,	*of those assembled,*
Auā ua afio mai le pa'ia	*Because of the sacredness of the*
maualuga o Au'va'a ma Aiga.	*presence of high chief Auva'a*
	and his family.
Ua afio mai le Ma'opū o	*The presence of the Glory of*
Nafanua.	*Nafanua.*
Ua afio mai le pa'ia o le Mātua	*The presence of the Majesty of*
o Lamositele.	*Lamositele.*
Ua susū mai Alo o Losina.	*The Sons of Losina are here.*
Ua maliu mai le paia o le	*The sacredness of the Four*
Toa'fā ma le fetalaiga a Silia	*Orators and of the orator Silia*
La'ei,	*La'ei is present,*
Ma le mamalu o le Tapua'iga.	*And the dignity of their*
	Assistants.

The men smiled, surprised that I could speak their esoteric rhetoric. Calls of *"Mālō!"* ("Well done!") were heard throughout the *fale.* But despite their praise, I knew that the real test of my knowledge still remained ahead. In contrast to the precisely articulated chiefly Samoan with which he greeted me, Fuiono Mase'ese'e now switched to colloquial language, with "k" sounds substituted for the "t," to tell a humorous story. Everyone laughed, and the jesting continued for some time, with everyone joining in the fun. After my entrance, the men ceased to pay any attention to me, and I became just part of the group, sitting in the hut and listening to Fuiono's story. For years I had longed to be so accepted in Samoan society that my presence, my foreignness, would no longer be noticed. In that little hut that morning, I glimpsed what it might be like to finally blend in, to have a clearly demarcated place in Samoan society.

In America and similar Western societies, wariness characterizes relationships between men. Although relative differences in physical, political, or financial power are seldom articulated, they are never ignored. A subtle accounting for differences in power creates a distance that makes it particularly difficult for men to form close bonds with coworkers after reaching adulthood. In such societies, which embrace egalitarianism in theory, but in fact defer to the strong, the rich, or the educated, a polite but detached demeanor proves prudent in most social settings—at least until one figures out for certain whose star is rising and whose is on the wane.

Such social ambiguity does not exist in Samoan society, which is perhaps why Western hierarchies appear so mysterious to many Samoans. In Samoa, one's position in the male echelon is always crystal clear: during kava ceremonies, the cup is passed from the highest chief to the lowest in strict order of rank. There is no point in jockeying for position, because status is strictly determined by one's chiefly title. The resultant hierarchy, frozen in the *fa'alupega,* can never change. Perhaps as a result of this immutability of social status, envy plays no significant role in chiefly councils. The Falealupo chiefs regard Fuiono's right to speak first as inalienable, and do not question it. Although such social stratification is foreign to American culture, steeped as we are in the cult of egalitarianism, Samoan stratification provides the comfort of always knowing where one is to sit, when one is to speak, and what one should say. And the Samoan social system produces in all a deep feeling of inclusion. Here, as a foreigner and an untitled man of Falealupo, sitting in a little hut, I felt as if I had been invited to fly in the corporate jet or to join the executives playing from the professional tee at Pebble Beach.

Our period of relaxation ended when a messenger arrived, summoning us to the large meeting house where the investiture ceremony was to occur. Each chief took his kava stick as he left the hut, and together we joined the stately procession to the investiture ceremony. As a group we entered the large, decorated *fale* and were given fragrant leis made from *moso'oi (Cananga odorata)* flowers. We shook hands with the four or five chiefs representing the families of those who were to receive the titles. The two chiefs-to-be were sitting on either end of the *fale,* gaily dressed in fine mats, beads, and paper currency folded in their hair. When the entrance rhetoric was complete, a young man dragged a mat before our group. As the mat arrived in front of each chief, he ceremonially placed his kava stick on it, a symbol of his respect for the families of those who would become chiefs. The young man paused in front of me, but I motioned him on. "You're supposed to throw your kava stick on the mat," the chief sitting next to me whispered.

"I know what I'm doing," I whispered back.

"All of the kava from Falealupo district has been received," the orator representing the prospective chiefs' families intoned. "Is there any other kava remaining?"

I slapped the mat in front of me and the young man dragged the mat around again. I placed my kava stick on it, making certain that the long stem faced toward the end of the hut.

"This kava is the respect from the distant country of America," I said in Samoan. "As a professor from the United States, I too, honor the candidates who wish to become chiefs."

There was a murmur of approval and broad smiles of surprise from the visiting chiefs. Lilo had taught me this little piece well.

The kava ceremony progressed through all of its intricacies. At the climax, the large tapa ribbons tying the fine mats around the candidates' waists were removed, much like unwrapping a Christmas present. Each was ceremonially served a cup of kava. As each partook, he became a chief, forever with a new name, title, and responsibility.

Lunch was served on banana leaves placed on woven trays. The high chiefs were served the *tuala,* the ribs of the pigs, while the orators were presented the *alaga,* or flanks. In addition, all were served taro and *palusami.* Conversation ceased while the chiefs quickly ate, a form of respect: no one can eat until the chiefs have finished. The chiefs then placed their considerable leftovers in baskets, which were carried away by members of their families. As soon as the meal was removed, hot Samoan cocoa was served.

After a few minutes of conversation, Fuiono Mase'ese'e cleared his throat and addressed the gathering. "I understand that our visitor from overseas, Koki, has something he wishes to say to the village council. Koki, the time is yours."

I slid slightly forward on the mat, and closed my eyes. In Samoan rhetoric, form matters far more than substance, and it is important to begin with composure. The orator who is most eloquent and persuasive will likely rule the day, regardless of the merits of his arguments. I knew the rules of the game—to the chiefs I must speak persuasively, without fear, without hesitation.

I began speaking slowly and loudly, in the precise intonation and cadence that Aumalosi taught me so many years before: *"E vae ane la le paia fa'atafafā o le maota lenei, ou te tau pa'i mālū atu ai i lo outou paia . . .* " "With due respect to the great sanctity of this house, I gently speak, because we are in the presence of the sacredness of Auva'a and Aiga. We are in the presence of the glory of Nafanua. We are in the presence of the majesty of Lamositele. The sons of Losina are here. We are in the presence of the Four Orators and their assistants. May the waters be calm, and may you be appeased because I have abused the sanctity of this gathering. I have no *agatonu* with you. But I praise God because we are one in Christ. Even though we are of different color, and speak a different language, it is because of His love that we are able to greet each other on this beautiful morning."

"I thank each of you because you have accepted me and my family, as well as my research associates, in Falealupo. It is a source of great pride to me that I am able to be part of the untitled men in this village. You have accepted me as an untitled man, and I have in my possession a fine mat named 'testimony,' which evidences that I am a true member of Falealupo village."

"In this morning gathering, I remember the words of Princess Leutogi Tupa'itea when she said that we have been saved in the crotch of a *Callophyllum* tree. Just as that woman was saved by the rain forest, today the rain forest continues to protect our lives."

"Two days ago, I saw the terrible destruction caused by the loggers. I know that you are not happy with that situation because the forest is precious to you. I have heard that you have for many years refused logging companies, but now have accepted their offer only because you need money to build a new school."

"Therefore, I gently approach your sanctity and sacredness with my humble opinion. I scratch the roots of the *fau* tree and beg your indulgence. Could I pay for your school so that you can save your rain forest? I have no other objective than the preservation of the rain forest. I do not wish to control your land nor to make decisions concerning your forest. I am merely proposing a covenant: I will raise money to pay for your school if you will protect your rain forest. I believe the forest is sacred because it was created by God's holy hand. We must find a way to save it."

"Thank you very much for the opportunity to visit you in clear skies and in health. I pray that God will cause the orbit of the moon to be high above the heads of the high chiefs. May God also bless the orators that their whisks may never fall nor their staffs ever break. May my voice continue to live. May our morning be blessed."

There was a moment of silence after my speech, and then the chiefs, along with the villagers assembled outside the hut, burst into

applause. Fuiono Senio exclaimed: *"O ia o le suli o Nafanua"*—
"He is the heir of Nafanua." Fuiono Mase'ese'e addressed me with
a short, formal reply:

"Thank you for your kind words. We all appreciate your excellent speech. As you know, anything involving land is very difficult
in our culture, so we will need time to consider your proposal.
Many thanks again for coming."

Fuiono motioned for me to leave, so I got up, shook all of the
chiefs' hands, and left.

"That was very impressive, Koki. Everyone was amazed at how
well you speak," Lilo said afterward.

"I didn't think you were coming," I said.

"I sat under the breadfruit trees behind the hut so I could listen
in. I was very proud of you."

"But do you think the village will accept my offer?"

"I don't know. Anything involving customary land in Samoa is
hard to predict. There will be much suspicion of your motives. We'll
just have to see."

The next three days seemed to last forever. Apparently the chiefs felt
that it was not appropriate to discuss my proposal immediately following the investiture ceremony honoring Foa'imea and Foa'imea,
the two men who had been granted identical titles, so they decided
to postpone public discussion until they could assemble the entire
village. The meeting was to include all of the chiefs, but I was very
conspicuously not invited.

I walked to Vaotupua in the southern part of the village to visit
healer Lemau's husband, Seumanutafa Siaosi. Seumanutafa, usually
jovial and full of merriment, was uncharacteristically withdrawn.

"I'm sorry, I really can't predict what will happen. Land issues
are complex in Samoa, and Samoans are filled with suspicion. In the
old days, the Germans gave some of the chiefs cigarettes, and had
them make a mark on a piece of paper. Without realizing it, the
chiefs signed away vast tracts of land for only a plug of tobacco. So
even though everyone in the village respects you, those memories
are still very painful."

"But don't you trust me? Can't you go and plead my case?"

"I trust you, but I'm in trouble with the village council over
another matter, and can't attend."

That day Bill, Dixie, and Thomas had driven around the island
to count flying foxes, so I did my botanical work alone. I avoided
the part of the forest that was being logged—it was just too upset-
ting to see again now—and collected medicinal plants in the littoral
forest by the sea instead. But I could not forget the scene of devas-

tation that I had witnessed. And I knew that hour by hour, more and more of the rain forest was disappearing.

That night I sat at the water's edge, and the waves swept over me. Why not let the waves carry me out? Perhaps I could disappear into forgetfulness of what had happened here, forgetfulness of how I had tried but almost certainly failed. Not far from where I sat was the westernmost point of Savaii—'O le Fāfā—the connection point in legend between this world and Pūlotu, the undersea world of spirits. Lilo had told me that Nafanua arose from the sea at the same spot.

"Lilo," I asked later that night, "Something that Fuiono Senio said at the chiefs' meeting puzzles me. Fuiono Senio called me the 'Suli o Nafanua.' What did he mean?" I had always found Fuiono rather intimidating: he struck me as the most aggressive member of the chiefs' council.

Lilo gave me a long, sideward glance. "Perhaps you should ask Lamositele tonight after dinner."

That night, sitting with Lamositele and Silia Tusi, an extremely brilliant orator, I posed my question again. The two looked at each other. Lamositele spoke first. "What Fuiono means is that you are animated by the Spirit of Nafanua."

"Why did he say that?" I asked.

Silia then spoke. "Koki, that is not a simple question. I must first help you understand who Nafanua was and what she stood for. And that will require some time to explain."

"I would be very interested."

Silia explained that Nafanua's father, Saveasi'uleo, was the god of both the sea and the underworld, and appeared as a combination of man and moray eel. Nafanua's mother, Tilafaigā, was one of a pair of Siamese twins, and her liaison with Saveasi'uleo resulted in a pregnancy. Saveasi'uleo routinely destroyed all of his offspring, eating them alive. Knowing Saveasi'uleo's viciousness, Nafanua's mother successfully concealed her pregnancy and childbirth, and hid the placenta deep in the ground. The child was thus named "Nānā Fanua"—"Hidden in the Earth."

But as a child, Nafanua once strayed too close to the beach. Suddenly the evil sea god rushed at her from the sea. Before he could seize her, however, her uncle Ulufanuasese'e surfed along the tops of the waves to decoy the monster beneath. Looking at Saveasi'uleo in ridicule, Ulufanuasese'e said: "Look what has become of you! Would you even kill and eat your own brother? We will separate: you stay in Pūlotu and I will stay on the land. But we will meet again at the end of time and our lineage." Saveasi'uleo slid back beneath the waves, but demanded the presence of his daughter in Pūlotu. There she remained under her father's tutelage and studied the art of destruction, but she also carried with her the knowledge of the rain forest and the healing power of plants.

On land in Falealupo, oppression reigned. The people were made slaves on their own land, forced even to climb coconut trees upside down. One day a man named Tāi'i called out in desperation: "Is there no one to save us?" Deep within the ocean, his words found an audience.

Saveasi'uleo commanded his daughter Nafanua: "Go up and free the people. Destroy the oppressors utterly with three war clubs: *Fa'auliulito, Ulimasao,* and *Tafesilafa'i.*"

Nafanua looked on the earth with compassion, however, and took only two of the war clubs, leaving behind *Fa'auliulitō,* lest the entire world be turned forever to ash. Nafanua swam toward the portal of the underworld, to the place of her birth, Falealupo.

Nafanua surprised the village with the ferocity of her solitary war against their oppressors. No one suspected that such a mighty warrior could be a woman until one day in Faia'ai village, the people were stunned to see her breasts. Chiefs came from throughout Samoa to pay her homage. Nafanua sought to establish for the first time a central government for all of Samoa headquartered in Falealupo, and redistributed all of the chiefs' titles. But to the villagers of Falealupo came some special charges, or *tofi.* "Auva'a" became a sitting monarch, and the priest of Nafanua. The rest of the high chiefs became her *aiga,* or family. One trusted orator she named "Fuiono" and charged him to be the spokesman for the village. "Taofinu'u" ("Hold fast the village") was charged with upholding the good of the village. "Soifuā" ("life") was charged always to protect the village's well-being. After she bestowed these titles, another prominent orator came running from the village to Nafanua. "I'm sorry, but I have no other *tofi* for you," Nafanua told him. "But look at your fine clothes. I will call you 'Silia laei' [literally, "beautiful clothing"], for you will always beautify the village with your presence."

Nafanua deigned that the village should be led, under Auva'a's direction, by the four paramount orators. The reign of Nafanua ushered in an era of solidarity and peace among the Samoan people. Temples were built to her, and Auva'a became her earthly representative. But before returning beneath the waves, Nafanua left a prophecy: "I have founded a government that will serve you well. But one day a kingdom will come from across the sea that is not of this earth, but of heaven. When it comes, you must enter it." It is largely because of this prophecy that, when John Williams of the

London Missionary Society introduced Christianity to Samoa, nearly all of the inhabitants of Samoa converted.

By the time Silia and Lamositele had finished telling me the story of Nafanua, men were starting to launch their canoes by lantern light for night fishing. It must have been nearly 2:00 A.M. Lamositele started to rise to prepare his own fishing gear.

"But Lamositele, I still don't understand. I think that is a beautiful story, but why would Fuiono say that I am filled with the spirit of this goddess?"

"Don't you see? Nafanua was not from Samoa. She just appeared out of the sea to fight our battles and save the village from oppression. She loved the rain forest and protected it. Well, we're now under oppression from this sawmill. We have nowhere else to turn to get funds for the school. Suddenly, you appear out of nowhere and want to kick the loggers out. You talk about loving the forest. Fuiono thinks that in some sense, Nafanua has returned."

I knew that the chiefs' discussion of my proposal would be difficult and protracted. Consensus is required for all major decisions. Absent consensus, there is simply no conclusion. As a result, important decisions can be delayed for weeks until everyone agrees. The morning after my proposal, all of the chiefs in Falealupo district met. Though the subject was perilous and difficult, consensus was reached in only a matter of hours.

It was afternoon when Lilo told me breathlessly of the chiefs' decision. "They accepted your offer! They have agreed to stop all logging if you can pay for the school!"

I offered a silent prayer of thanks. I later learned that Fuiono Senio, Tāi'i, Soifuā, and Auva'a all championed my proposal, but that Taofinu'u regarded it with suspicion. What would happen after I paid for the school, he wanted to know. Would I gain control of the village lands? Why would I want to pay for a school and expect nothing in return? Surely I must have something up my sleeve.

Tāi'i argued that God must have sent me to deliver Falealupo from its terrible dilemma. Auva'a pointed out that the logging company might cut down the whole forest, and there still wouldn't be enough to pay the debt for the school.

Fuiono Senio, sensing that a consensus could be reached, asked Taofinu'u if he couldn't join his fellow chiefs in accepting the proposal. Taofinu'u, seeing that he was the only holdout, agreed that the village had little recourse if it wished to preserve the forest. They would all travel down this new and perilous road together.

Fuiono and the other three paramount orators asked to meet with me so that they could discuss my offer in detail. I told them that I would go to the Development Bank in Apia and immediately have the mortgage on the school signed over to me as a personal obligation. I had arranged for mortgage payments for six months, during which time I would raise in the United States the funds necessary to pay off the mortgage completely. I planned to return to negotiate a covenant protecting the forest forever in return for paying for the school. I would formally renounce any rights to the village lands or forest and would negotiate the covenant completely in the Samoan language with the village. All I asked in the meantime was a complete moratorium on all logging and flying fox hunting during the six-month period.

Fuiono nodded agreement. I spoke again. "I know that this has required a tremendous leap of faith for you and thank you from the bottom of my heart. I will not let you down." As I left, Fuiono muttered, "He is the spirit of Nafanua."

The chiefs' decision to protect the forest rapidly spread through the village, and village representatives informed the loggers. However, the loggers apparently did not believe that such a reversal was possible, and showed up the next day for logging as usual.

Word that the loggers were still cutting the forest in direct opposition to the village decision reached Fuiono Senio, who was visiting friends by the sea. Fuiono ran three miles to the logging site with his machete in hand. Motioning the bulldozers and saws to stop, Fuiono spoke to the astonished group of loggers.

"Don't cut another tree! This forest is now taboo!"

Apparently one of the loggers didn't believe Fuiono and reached for his saw to continue working.

"Don't touch another tree! Try it and watch what happens! You'll turn to dust!"

The head of the logging party spoke respectfully to Chief Fuiono, well aware that any insult given to a Samoan chief in his own village could lead to immediate violence.

"Excuse me, sir, but is there a reason why we should stop cutting these trees?"

Fuiono replied vehemently. "Yes. Let me explain it to you. This forest is now taboo. I am the chairman of the village forestry committee and have authority over the entire forest. And the chiefs' council has decided to taboo the forest. You are all now standing on taboo ground. You should all leave immediately."

The logging company employees, all Samoan, looked at each other. They had heard words that have deep saliency in Samoa:

taboo, authority, chiefs' council. To willingly violate taboo is an unthinkable offense. Besides, they had heard Fuiono threaten one of their own workers. They knew that even with all of their machinery, they stood little chance against an entire village. This was clearly a problem for the front office to sort out. Quietly they packed up their equipment and left.

10 cm.

1 cm.

Piper graefü

CHAPTER 9

Dreams

*A man that is born falls into a
dream like a man who falls into
the sea. If he tries to climb out into
the air as inexperienced people
endeavour to do, he drowns. . . .
The way is to the destructive
element submit yourself, and with
the exertions of your hands and
feet in the water, make the deep,
deep sea keep you up.*

Joseph Conrad, *Lord Jim*

I sat alone on the white bench in the Development
Bank office above Coxon's store in Apia. Chiefs of
various villages, all with leather cases, clustered in
the waiting area. Soon a middle-aged Samoan woman addressed me in English with a slight New Zealand accent.
"Have you been helped?"

I responded quietly in Samoan. "I need to see the bank director.
I wish to assume the mortgage for the Falealupo school."

The bank employee looked at me in astonishment. "You speak
excellent Samoan. What is your name again and what do you
want?"

When I detailed my business, her eyes widened. "You mean you
want to pay their school loan, just so they can save the forest, and
don't want anything in return?"

"Right."

The woman reverted to English. "Mr. Cox, your name in
Samoan should be *Agelu*—because I think you are a real angel. Let
me see if Mr. Meredith, the bank manager, can see you."

Rudy Meredith, the director of the Development Bank of Western Samoa, is a well-built, handsome Samoan. He spoke to me in
perfect English.

"Good morning, Mr. Cox. I've heard of you. Aren't you friends with Dan Betham?"

"Dan and Kathy Betham are like my adoptive parents."

"Dan is on our board of directors and we have the utmost respect for him. Can I have some tea or coffee brought in?"

"No, thank you."

"So, you really want to assume this school loan yourself?" he asked, opening a large brown folder a bank clerk brought in.

"Yes. I need to set up some method of transferring funds by wire, so I can make regular payments for six months. At the end of that time I plan to pay off the mortgage completely."

"Are you aware that the balance is approximately $65,000 U.S.?"

"Yes."

"And do you know that the village has already been advanced as much as $20,000 U.S. by the logging company in addition?"

"No, but I will pay that too."

"Mr. Cox, we will be pleased to assist you in any way that we can."

I signed the necessary papers to place the loan in my name, and wrote down the wire transfer information. Walking down the steps of the bank building, I caught a cab to the airline office to book a return flight to the United States. My research team, travelling from Falealupo, would fly later. I would need as much time as possible in the United States to raise the necessary funds.

Leaving the airline office, I decided to indulge in what had become for me a little ritual I enjoy before leaving the country: I went to the open-air market in Apia. On entering the market I was assaulted by the familiar cacophony of sound and myriad sights and smells. Long rows of vendors, sprawled either directly on the rough concrete floor or seated on woven mats, displayed their taro, pine-apples, and sugarcane on low tables. The press of large Samoan bodies bedecked in garish floral patterns was both intoxicating and suffocating. I walked past two rows of white *Tridacna* shells, some as large as my forearm, and a row of conches, pink and gleaming. The vendors also sold scented Samoan coconut oil, dispensed in old Coke bottles sealed with coconut corks. There was a row of fist-sized blocks of raw cocoa. I brushed past displays of white *pule* shell leis, whisk brooms made from coconut leaf midribs, and coils of sennit woven from coconut fiber. On the ground were baskets of a medicinal plant, *fue manogi (Piper graefii)*, a rain forest vine used to treat a variety of ills.

Next were the fruit sellers. Ripe papaya cost fifty cents a basket; bananas ranged from the tiny but sweet *Misi Luki* to the gigantic

Fa'i Samoa that are buried in the soil to ripen. Oranges, green when ripe, were stacked in neat pyramids. Other tropical fruits also abounded: mangoes, mountain apples, star fruit, custard apples, and chestnut-like *ifi* that are roasted before eating. Behind the aisles of fruit were small pantries offering plates of cooked taro, rice, fish heads, mutton stew, coconut buns called *panikeke,* and the occasional pig leg.

In the market's northwestern corner are sold fresh *umu*—hot foods from Samoan stone ovens, proffered in freshly woven coconut baskets. Here you can buy hot breadfruit, taro, *palusami,* lobsters, packets of reef fish, and a variety of Samoan marine delicacies. An acid test of Samoan acculturation for any *palagi* is the ability to eat *sea,* the expelled intestines of a marine coelenterate. These invertebrates are collected at low tide, and their intestines are sucked out by women and then spit into old whiskey bottles, where they are fermented in the sun for several weeks. The result: a mucus-like product that is bitter, rancid, and yellow-green with effervescent bubbles that fizz like a carbonated drink shaken by a child.

At the west side were sellers of *mea mata:* uncooked taro, green bananas, and large *ta'amu* rhizomes. These Samoan staples are displayed in small heaps directly on the concrete floor. Tourists seem intimidated both by the heaps of strange produce and by the coarse appearance of the vendors, and veer away from this part of the market. But in their decision to engage the more Westernized sellers of trinkets instead, they miss some of the best produce of Samoa and some wonderful human beings.

Over the years, I have made several friends among the *mea mata* sellers. I was particularly fond of Taimani and hoped to spot him. He, his wife, and their two small children lived in a small tin-roofed hut perched on stilts near the Tīavī waterfalls high in the mountains of Upolu. I had spent several delightful afternoons and evenings there with them while I botanized nearby. The rich volcanic soil in their region produces some of the best taro in Samoa. To supplement this starchy diet, Taimani hunted flying foxes and fruit pigeons. But as he witnessed the dramatic decline of flying fox populations throughout the islands, he eventually gave up hunting them entirely.

Taimani was about six feet tall, with narrow hips and broad shoulders. The scars on his face evidenced his former attempts as a prizefighter; on his arms were garish tattoos of women, stars, and a serpent. Across the top of his hand was crudely etched a date—Sept. 23, 1965—commemorating neither his birth nor anniversary, but merely the date his hand had been disfigured with the tattoo.

Taimani had dropped out of elementary school, and spoke in rough, colloquial Samoan. His young wife, slender and handsome, addressed me formally, greeting me in the best respect language she knew. She always seemed to have hot taro, mutton flap stew, and dark Samoan cocoa ready whenever I dropped by. We would chat in their hut, overlooking their taro plantation and the southern Upolu coastline far below.

Taimani was an alcoholic, but lacking cash to buy beer, he became an expert at *fa'amafu*—home brew—and occasionally hosted raucous gatherings in his hut. It was at one such gathering that Taimani got into a fight. In a drunken rage, his guest attacked him with a machete. Taimani grabbed his shotgun and fired a shot from the hip. The man died from his wounds. Taimani was convicted of manslaughter and confined in Tafa'igata prison. His wife, all of twenty-eight years old at the time, was left alone with their two daughters to tend their taro patch.

I learned of the tragedy and visited her in her hut. Each Saturday, she told me, she would carry the heavy burlap bags of taro to the roadside, flag down a truck, and pay the driver to take her and her taro to the Apia market. After selling her produce, she would buy sugar, salt, and several cans of mackerel for the children, and return by bus to the saddle road near her hut.

I visited Taimani's family several times during that period. His wife protested when she discovered the small amount of cash I left behind on the mat, but I told her it was for the children. When Taimani's sentence ended, he hugged me with his rough, tattooed arms. Since then, I have always tried to see him when I visit the *mea mata* sellers in the market.

But today, I couldn't find Taimani. I was ready to leave the market when I heard his wife's voice call to me through the crowd. I turned and saw her motion to her red blouse. There, held carefully against her chest, was an emaciated infant *Pteropus samoensis*, the endangered Samoan flying fox.

"Where did you get it?" I asked.

"The hunters came and killed the flying foxes. But we found a dead female that they lost in the bushes. Nuzzling against her breast, trying to nurse, was this little baby."

"How long have you had it?" I asked.

"I found it a week ago and kept it alive by feeding it crackers and soda pop. Here—it's for you. I knew you would come."

She handed me the little flying fox and the remains of a roll of imported tea biscuits. Very obviously Taimani's wife had been spending more money on the care of the infant bat than on her own two children. I pulled out a twenty *tālā* bill (about $10 U.S.) and handed it to her.

"I can't take this," she said. "It's a gift. Just for you."

"No, please, take the money. Just this once. That's my culture. I've got to pay you so that no one can ever question whether I truly own this creature."

Puzzled, she took the money and bade me farewell.

I cradled the little bat and took it to Aggie Grey's Hotel. Although Thomas and his family had left for Sweden, Bill and Dixie planned to stay at the hotel while they made some surveys in Upolu. While Bill and Dixie tended the infant bat, I ran out to buy some human baby formula and a syringe. Returning to the hotel room, I mixed the baby formula and filled the syringe. Our first attempts to inject milk into the bat's little mouth resulted in a bit of coughing, but soon it learned how to drink from the syringe, and emitted a contented gurgle.

"The poor thing," Dixie murmured, nuzzling the little flying fox against her cheek. "What should we call him?"

"*Pe'a vao*," I answered, thinking of Michael Rothman's sing-song call. "The Samoan name for *Pteropus samoensis*."

Not knowing if the little flying fox would live through the night, Bill and I rigged a makeshift perch for him to hang from, and then measured and photographed him from every possible angle.

"How am I going to get him back to the United States?" I asked.

"That is a nontrivial question," Bill answered. "First you need a U.S. Fish and Wildlife Permit for Injurious or Dangerous Animals."

"What's dangerous about flying foxes?" I asked.

"It seems that fifty years ago, someone decided that flying foxes posed a potential menace to California agriculture. So they put them on the list of dangerous and injurious wildlife. Importing this little guy into the United States is considered tantamount to importing a live cobra."

I spent the next few days securing permission from the Samoan government to export the flying fox. Eddie Thompson, chief quarantine officer, issued me an export permit. Meanwhile, Bill had been telephoning and sending faxes to the Fish and Wildlife Service in Washington, D.C., but true to form, the officials there were unhelpful. It would take a minimum of six weeks to process the paperwork, Bill was told, and even then it might not be issued. We didn't have six weeks: the animal would die if we were unable to care for it in the United States. Rules were rules, the Fish and Wildlife Service replied.

I had a sudden inspiration. I contacted our friend Carol Lambert, who served as an aide to Congressman Howard Nielsen from our home state of Utah. Perhaps a call from a U.S. congressman might spur some activity. I then called a friend in Pago Pago—Eni Faleomavaega, the lieutenant governor of American Samoa.

The next morning, a message came back from the governor's office in American Samoa—my permit to import the bat to Pago Pago would be waiting for me at the airport. While Bill and Dixie flew on to the United States, I stayed with Eni at the lieutenant governor's mansion and took more photographs of little Pe'a Vao eating a variety of Samoan fruits. It took three days for the U.S. Fish and Wildlife permit to arrive. "It's odd that they dragged their feet when you called," Carol told me on the phone. "They seemed so helpful when Congressman Nielsen chatted with them."

Carol told me that I would be met in Honolulu by U.S. Fish and Wildlife agents as well as by representatives of the U.S. Center for Disease Control and the Hawaii State Agriculture Department.

"Why all the muscle?" I asked.

"They want to physically escort you to your plane to Utah," she said. "They don't want the flying fox getting loose in Hawaii."

I had covered all the bases, or so I thought. It was only when I arrived at the airport that I realized that I hadn't obtained permission from Hawaiian Airlines. I didn't dare chance it at this late date. Eni arranged for me to wait for the flight in the VIP lounge at the Pago Pago airport. So prior to my flight, the little flying fox happily cavorted about the furniture normally used by visiting dignitaries. Just before leaving the lounge, I slipped Pe'a Vao headfirst into a cotton stocking. "Sorry about that, but I don't want the stewardesses to see you," I whispered.

On the aircraft I was seated next to a stocky but pleasant Samoan man. After we had taken off, he noticed the sock wriggling in front of my seat.

"What have you got in there, a rat?" he asked.

"No—it's a baby flying fox. But please don't let the stewardess know. *Palagi* are frightened by bats."

"Well, at least let me see him," the Samoan implored.

I waited until the flight attendant had gone back to the galley, and then pulled out the infant flying fox. The man giggled with delight. "Well, isn't it cute," he said, tickling the bat with his chubby finger. The woman seated behind us overheard us and peered across the seat. "They've got a live flying fox!" she exclaimed in Samoan. Soon calls of "Hold it up!" and "Let us see it!" began to ring out in Samoan.

I turned to face the rest of the passengers in the aircraft. "Please help me," I said in Samoan. "I'm taking this baby flying fox to America to be cared for. All the governments involved have approved, but I haven't secured permission from the airline. *Palagi* are afraid of bats, and if the stewardess finds out, I might get in trouble."

"We won't tell anybody!" the Samoans said cheerfully. "But at least let us see it!"

Slowly I lifted up my arm, on which the tiny flying fox hung. A cheer and hearty applause burst forth from the Samoan passengers. Hearing the commotion, the flight attendant came running from the galley. I dropped back into my seat and quickly draped an airline blanket over my arm. The Samoans suddenly turned stone-faced. Puzzled, the flight attendant returned to the galley. I stood up again with the flying fox, and the look of delight was evident on the Samoans' faces. One woman, and then another, started passing up to me the fruit portions of their in-flight meals.

"The poor little thing is probably hungry," one woman said as she passed me her bowl of fruit cocktail. And so the rest of the flight progressed—the flight attendant puzzled by occasional outbursts of applause and laughter from the passengers, and the Samoans and I all conspiratorially engaged in a grand but delightful smuggling operation.

When the plane landed in Honolulu, I was met at the door by two polite but serious federal wildlife agents in uniform and an officer of the Hawaii State Agriculture Department. They examined my passport and importation permit, and then escorted me through a back door to the waiting Delta flight to Salt Lake City. Once I had boarded the aircraft, they retreated.

I had bought a first-class seat from Honolulu to Salt Lake City, guessing that the airline might be more willing to tolerate eccentricity from a front-cabin passenger. But five minutes before the aircraft was due to take off, the well-dressed gentleman seated next to me looked at the squirming sock at my feet.

"What have you got in there, a rat?"

My mind raced. If the airline threw me off the flight and back into the arms of the Honolulu Fish and Wildlife agents, there was no telling what would happen to the flying fox. My activism on behalf of the animals had not particularly endeared me to the Honolulu regional office of the U.S. Fish and Wildlife Service. It had been made abundantly clear to me as well that nobody messes with the Hawaii State Agriculture Department. I suspected that there was a good chance that even a U.S. congressman could not save my bat from destruction if we were forced to remain overnight in Honolulu. But I decided I had to tell the truth to my fellow passenger.

"I don't want to alarm you, but I am carrying an infant flying fox in the sock," I said.

"Let me see it," the man asked.

"I'd prefer not to show it to you until we've taken off. It might alarm some of the other passengers, and the airline doesn't know that I have it. While they can easily throw me off the aircraft before we take off, I don't think they will be as willing to dump a full load of fuel to return me to Honolulu once we're airborne."

"Don't worry," the man said. "I'm an off-duty Delta pilot, and I won't let anything happen to you or your . . . flying fox. But I do want to see it after we take off."

After we were in the air and the captain had turned off the seat belt sign, I pulled Pe'a Vao from his sock and nursed him with the syringe.

"Gosh—he's cute!" the pilot exclaimed. "Do you mind if I take him up to the cockpit to show the captain?"

"No, I'd be happy to take him up there," I replied.

"I'm sorry, but FAA regulations forbid passengers from entering the cockpit during flight. But this flying fox—well, I reckon he's a fellow aviator."

The off-duty pilot gently hung the flying fox from his arm and went to the cockpit. A few minutes later he returned.

"They sure liked him up there," he said. "The little guy flapped his wings just like he knew we were in the air."

By this time the other first-class passengers had seen the flying fox, and several children asked if they could be photographed with him. Word spread among the cabin crew, and they started trickling up from the coach cabin to see their unusual passenger.

"He's darling!" a flight attendant exclaimed. "I never knew that bats could be so cute!"

My children were all at the Salt Lake City airport to greet me, and we quickly hustled Pe'a Vao into the warm car. Barbara set up a wooden clothes drier in the anteroom of our house. Pe'a seemed happy to hang from it, and quickly fell asleep after his long journey. It was wonderful to be home and to have brought Pe'a back successfully, but I felt a bit depressed. Although I was relieved that the rain forest had been given a temporary reprieve, I had signed the mortgage, and now had to come up with the money. And I had only a short time to do it.

After putting the children to bed, Barbara and I discussed the worst-case scenario. If we sold our house, most of our possessions, and our few investments, we would still probably run short of the required $85,000. But Barbara was resolute in her determination to help the villagers. "It's worth it," she said. "How many times in our life will we get to save a rain forest?"

The next day, Barbara and I took the children to a movie—*Bambi*. In the film, the primeval forest is destroyed by fire, and hunters kill Bambi's mother. I was embarrassed by my emotional response to a cartoon, but turned to see that Barbara and the children were equally distraught: the film was all too reminiscent of our experience in Samoa.

That night, I still couldn't sleep. Whenever I closed my eyes, the faces of the village children would appear. I imagined too the gentle countenance of Pela, the distinguished look of Lamositele, and the beautiful village nestled on the white sand beach. I thought of the trusting brown eyes of the infant Samoan flying fox that was sleeping peacefully just down the hall. And then I could see the crash of tree after giant tree falling to the ground, trailing a stream of lianas like the limp nylon cords of a failed parachute.

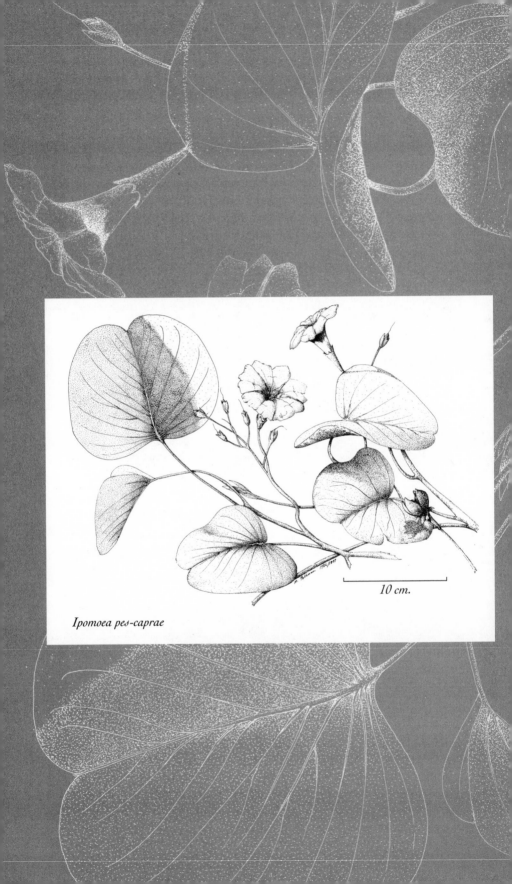

Ipomoea pes-caprae

10 cm.

CHAPTER 10

Responsibility

In dreams begins responsibility.

W. B. Yeats,
Responsibilities

ashington, D.C., in August can be as hot and muggy as Samoa. But the sounds from the street that day in 1988 made it clear that I was no longer in the islands. Verne Read's brother-in-law, Chapman Chester, had lent us his Dupont Circle condominium. Verne, along with Merlin Tuttle of Bat Conservation International, had flown in the previous day, as had Governor Lutali and two Samoan chiefs, Tua and Tuaolo. Michael Rothman had come by train from his home in Manhattan. All of us had been scheduled to testify before the House Subcommittee on Public Lands and National Parks on the proposed Samoa National Park. Since the fate of the Samoan rain forest was being decided, I thought it also appropriate for Congress to hear directly from an endangered species that occupied it—Pe'a Vao, who once again had passed unnoticed through airport security on my way from Salt Lake City. Merlin, always a media whiz, had put the word out to CNN, local television and newspapers, and several reporters for the wire services.

Television cameras and reporters were waiting when we arrived at the ornate House caucus room the next morning. Verne and Merlin were dressed in expensive-looking suits; I wore the best suit

I owned. Governor Lutali wore a jacket and a formal lavalava, and Tua and Tuaolo wore only bark-cloth lavalavas and orators' whisks. Pe'a Vao seemed fascinated with what was happening.

Congressman Bruce Vento of Minnesota and the other members of the House Subcommittee on Public Lands and National Parks were already seated. Congressman Vento gave a warm welcome to the distinguished visitors from Samoa, and much to their surprise, formally presented them with a fine mat in accordance with Samoan custom, after which he gave me an appreciative wink. With Pe'a looking on from his mesh cage, the two orators stood with their staffs and spoke to the assembled congressmen in Samoan as I interpreted for them. I could tell that the congressmen were stirred by their eloquent pleas for the creation of a U.S. national park in American Samoa. Verne, Merlin, and I then testified. We spoke in favor of the proposal, which would protect about eleven thousand acres of rain forest on Tutuila island (including the Afono ridge where we had studied flying foxes) and Ta'ū island. The land would be acquired through fifty-five-year leases from the chiefs and villagers, who would be allowed to harvest medicinal plants, grow crops, and fish on park lands using traditional techniques and tools. All signs and interpretation in the park would be bilingual, in Samoan and English. Since the park lands were mountainous and of limited value for agriculture and forestry, there was little local opposition to the park proposal.

We were followed by an Administration witness representing the U.S. National Park Service. Since President Reagan had not shown great interest in expanding the National Park system, we were all concerned about what the official might say, but thankfully, he expressed neutrality vis-à-vis the proposed park.

After the meeting, all of the congressmen wanted their photos taken with Pe'a Vao, after which I returned him to Verne's brother-in-law's apartment. Later, I met the head staffer for the subcommittee, Dale Crane, for lunch in the Library of Congress. Things had gone better than expected, Dale explained. He was particularly impressed with the bipartisan support for the park, and was surprised by the strong testimony given by Congressman Jim Hansen (R-Utah) in favor of the proposal. Such a level of political support was sufficient for one of two possible additions to the park to be included in the final bill that would be sent to the full House. Which would it be—the Malotā rain forest on the eastern side of Tutuila, or the coral reef on Ofu island in the Manu'a group?

Landowners in both areas had expressed strong support of the park, I responded. Couldn't we have both additions?

"No."

"How long do I have to decide?" I asked.

"By the time we finish lunch."

The Malotā forest borders a beautiful piece of coastline and contains the largest roost of white-necked flying foxes, *Pteropus tonganus,* on the island. It also is the only forest in American Samoa where the rare Samoan palm *Clinostigma samoense* can be found. But the beautiful coral reef of Ofu, next to a gorgeous white beach adorned with beach morning glory *(Ipomoea pes-caprae),* could be seriously harmed by tourists traveling to Ta'ū to visit the national park. And the Ofu reef is perhaps the most beautiful and varied in all of Samoa. At a traditional luncheon held in Ofu during congressional field hearings, the chiefs and orators made a plea that their island be included in the new park. Although I longed to save additional rain forest, I realized that marine biodiversity is just as precious.

"We'll take the reef," I said to Dale.

That afternoon, I met with representatives of National Geographic Television who were interested in making a film on the Falealupo project. When I expressed concern at the idea of a large American film crew disrupting the village, they offered to send a husband-and-wife team. But they declined my request that they film all of the villagers speaking Samoan. They explained that American audiences don't respond well to subtitles. So I instead decided to accept an offer from Swedish filmmaker Bo Landin to produce a documentary on the Falealupo rain forest.

While I was in Washington, I visited Dr. Gordon Cragg at the National Cancer Institute. I brought slides of my work in Samoa and presented a seminar at the institute on Samoan herbal medicine and the use of *Homalanthus nutans* by the healers to treat hepatitis. Afterward I relaxed in Gordon's office.

"We're excited about the activity we're getting from your *Homalanthus* sample," Gordon said. "As I told you in our letter, it's extremely potent in our assay. We're using the antiviral drug AZT as the control."

He showed me the graph of the dosage response of sample 842. It looked a lot stronger than the AZT control, with nearly complete protection of human cells long before the dosage became toxic enough to inflict damage of its own.

"John Cardellina at Michael Boyd's lab is trying to isolate the responsible compound," Gordon said. "But we need more material. Could you get us several more kilos of this plant?"

"Sure. But there's one problem. Loggers started cutting down the forest that I collected sample 842 from. The village has temporarily stopped the loggers, but for the forest to be preserved, I have to raise enough money to build a school for the village, so that the villagers are no longer dependent on the loggers."

"Good grief!" Gordon said. "Is that the only place the tree grows?"

"No, it occurs elsewhere in Samoa. But that tract is one of the largest remaining rain forests on Savaii. I think it's important to collect from the original source, as plants collected from other populations might differ in their chemical constituents."

Gordon nodded. "Paul," he said as I was leaving, "Don't tell anyone about the activity of *Homalanthus*. We don't want to have the patent trail clouded. This may be an important discovery."

Under the bipartisan leadership of Congressmen Bruce Vento and Robert Lagomarsino, House Resolution 4818 establishing The National Park of American Samoa was passed unanimously. We suspected that the Senate might be harder sledding. After careful discussions, we arranged for Senators George Mitchell (D-Maine) and Orrin Hatch (R-Utah) to sponsor the legislation. But as the Senate's term drew to a close that September, our bill was placed in a large omnibus package of land legislation. Because of their opposition to an Idaho land exchange, the Sierra Club lobbied vigorously against the whole package. Discouraged, I left the United States on the last day of the Senate's term for a quick botanical expedition to Africa. En route to Nairobi, I called Barbara from a pay phone at London's Heathrow Airport while I was changing planes. She was so excited she could hardly talk. "Paul, the National Park bill passed the Senate."

I was dumbfounded. Apparently, in the last hour before the Senate recess, Senator Hatch pulled the park bill out of the omnibus package, and it was unanimously passed on a voice vote. Now the trick would be convincing the president to sign it. During his two terms, President Reagan had added only one unit to the National Park system, Great Basin, largely as a favor to his friend Senator Paul Laxalt from Nevada. How could I influence the presidential pen from Africa?

Quickly finishing my study of sea grass pollination in Kenya, I caught a flight to Zimbabwe, where I had planned research on the rare aquatic plant *Lagarosiphon*. I upgraded my accommodation to a hotel with IDD—international direct dialing. Over the next forty-eight hours, I racked up several hundred dollars in phone bills. My first call went to Steve White, a friend who lay dying of cancer in California. Steve was vice president of Bechtel and a close personal friend of U.S. Secretary of State George Schultz. He readily agreed to ask Secretary Schultz to discuss the importance of the park with the president. My next call went to former Harvard classmate

Roger Porter, who served as chief economic adviser to the president. The National Park of American Samoa would protect some of the very few rain forests that occur on U.S. soil, I said. Would he brief White House chief of staff James Baker on the issue, with an eye to having Mr. Baker ask the president to sign? The third call went to Senator Hatch's office. After expressing gratitude to his staff for the passage of the Senate bill, I asked if the senator could make a personal visit to the president. Senator's Hatch's staff said they would try to arrange it.

My fourth call was to 'O le Vao Matua, the rain forest conservation organization in Pago Pago. Barbara and I had anonymously sponsored a coloring contest for schoolchildren in American Samoa, with 'O le Vao Matua awarding prizes for the best drawings of what a future national park would look like. Armed with the White House fax number, Vao Matua faxed two hundred children's drawings to President Reagan. The president was known to have a soft spot for children.

Meanwhile, Barbara and congressional staffer Carol Lambert called every single U.S. senator and representative they could, asking them to urge the president to sign. Verne and Merlin made similar efforts, and in Pago Pago, Governor A. P. Lutali and Lieutenant Governor Faleomavaega persuaded Hawaii Congressman Daniel Akaka and other legislators to contact the White House. The National Parks and Conservation Association organized support, and BYU students sponsored a write-in campaign. I was ecstatic when President Reagan finally signed on October 31, 1988. But our work was unfinished: the Falealupo forest was still at risk.

At Brigham Young University in Utah, where I taught freshman biology, the students voted me "Professor of the Year" and asked me to speak on a topic of my choice to the entire student body. That month I also received a request to present a command lecture on my Samoan research for King Carl XVI and Queen Silvia of Sweden. Using the BYU presentation as a practice run, I made a play on the title of Lasse Hallström's popular Swedish film, *Mit liv som hund,* "My Life as a Dog." For both audiences, I spoke on *Mit liv som flygande hund,* "My Life as a Flying Fox," describing efforts to save the Samoan rain forest, culture, and flying foxes. Although I did not solicit funds, at the end of the Utah lecture numerous students made cash donations, and at the gala dinner after my lecture in Stockholm a spontaneous collection was made, topped the following day by a donation of several thousand dollars from Queen Silvia.

I was concerned about proper accounting for the donations, and received permission from BYU President Jeffrey Holland to set up a donor account at the university. But although these contributions were certainly a splendid start, it was just as certain that they would fall far short of the needed $85,000. I needed some major donors and I needed them fast.

Verne and Marion Read would undoubtedly have been willing to make a major contribution, but I could not in good conscience ask them for anything more. Verne had already fronted the mortgage payments for six months, and had invested a great deal of effort in successfully lobbying Congress for the National Park. He was also a major benefactor of Bat Conservation International, and Marion was on the International Board of the Girl Scouts. But this decision did not help me sleep. Each day, after teaching my university classes, I returned to our home just east of campus—the home that we would likely be forced to sell if I couldn't come up with the needed funds.

One evening after school, I heard some familiar laughter on my front step. Jim Winegar and Dan Wakefield, former Mormon missionaries to Samoa and leaders of numerous tours to the islands, were on my doorstep. They called themselves the "Wild Ducks," after the Samoan proverb *"E lele le toloa ae ma'au i le vaivai"* ("The wild duck flies but always longs for its nesting ground"). Jim and Dan regarded Samoa as their nesting ground, and always found some way to get back there every year. They had heard of my efforts to protect the forest.

"Do you know Ken Murdock?" Jim asked. "He might be willing to help out."

Ken owned Nature's Way, a large herbal medicine company, and had also served as a missionary in Samoa. Dan and Jim offered to make an appointment for me.

Judging by the size of the employee parking lot at Nature's Way in Springville, Utah, I guessed that perhaps six hundred people worked there. Jim and I had been waiting a short time in the thickly carpeted conference room when Ken Murdock and a man he introduced as his head attorney, Loren Israelson, greeted us. Ken wore an open sports shirt and khaki trousers. He had the build of a college athlete and could easily have had a career modeling sportswear. He grasped my hand firmly and gave me a warm smile. Loren was dressed in a three-piece suit. He was pleasant, competent, and, I suspected, there to protect his boss from making any hasty decisions.

Ken turned to me. "Well, Paul, what can I do for you?"

Sensing that Ken liked directness, I got straight to the point. "I'd like to leave here today with your $25,000 check to help build a school in Falealupo, Samoa, so that we can save a rain forest."

Ken smiled. "And what do I get for my $25,000?"

"Absolutely nothing. In fact, as a condition of donation, I will require you to renounce any rights to the school, land, or forest of Falealupo."

Loren looked horrified, but Ken didn't miss a beat. "Well, it's not every day I get an offer like that. Please, tell me more."

I recounted to Ken the plight of the Falealupo rain forest, and could tell that he was deeply moved. During his three years in Samoa he had developed a particular love for the poor people living in the outer islands. "Count me in," Ken said when I had finished.

I practically floated home. The second installment of the answer to my prayers came scarcely two weeks later when I attended the annual Samoan missionary reunion in Salt Lake City. These reunions are happy occasions, filled with Samoans visiting the annual L.D.S. church conference and former *palagi* missionaries pleased to practice their language skills. Island delicacies are flown in from Hawaii and California for the occasion. During the festivities, a striking man with gray hair and deep green eyes stopped me. "You're Paul Cox, aren't you? I'm Rex Maughan. I hear you've been trying to save a rain forest down in Samoa."

I explained the situation in Falealupo and the need to raise funds to build the school. He studied me for a few moments. "I'd like to help you. How much do you need?" he asked, reaching inside his suit coat pocket for a checkbook.

I stumbled. "Well, anything would help. It works out to a little over two dollars an acre. So whatever you can give would be appreciated."

"Would $30,000 be of help?" he quietly asked.

I tried to stammer out a "thank you." Rex put his arm on my shoulder and looked into my eyes. "Paul, I'm happy to support what you're doing. Please stay in touch."

As Rex Maughan moved away to mingle with others at the reunion, I sat down hard on a bench, completely stunned. In a minute or two, Jim and Dan came forward with big grins on their faces.

"We hear you met Rex Maughan," they said with a twinkle in their eyes.

I was speechless.

"We served our missions together back in '54. He always had a special place in his heart for Savaii," Jim said. "He owns Forever Living Products as well as a chain of hotels in Western national parks. He is the largest manufacturer of aloe vera products in the country."

I could hardly drive home as I thought of my good fortune. Combining the donations of Ken Murdock and Rex Maughan, we were within $10,000 of having enough money to pay the $65,000 bank debt and could use the Swedish funds toward repaying the logging company. If we could raise an additional $10,000, Barbara and I could use our own savings to satisfy the logging debt, and we wouldn't have to sell our house after all.

As I thought about how to raise the remaining $10,000, I considered the principle of tithing. If Mormons and members of some other denominations were willing to donate a tenth of their incomes to charity, perhaps some businesses might be willing to donate a portion of their profits to rain forest conservation. I thought through the types of businesses that might have ties to Polynesia. The major export of Samoa is copra, and I tried to think of any contacts in the coconut-oil business, but was stumped. Samoans also export a very high grade of cocoa, but again I was stymied: I couldn't think of a personal connection to a chocolate company. My mind began to wander across the centuries of European interaction with Polynesia. Although Polynesian culture had played such a large role in the European mind, the first European explorers left very little to the Polynesians other than decimating diseases. It all began with Captain James Cook's voyages in the 1700s to Tahiti and beyond . . .

As soon as I thought of that remarkable explorer, I knew I'd hit on the answer.

The telephone line to London was remarkably clear, although it took some time to get through the switchboard of the British Museum of Natural History to Chris Humphries. One of the world's top plant taxonomists, Dr. Humphries has been a pioneer in developing new computer-intensive methods of organizing plants and animals into natural groupings that reflect their evolutionary history. This field, cladistics, exploded onto the scientific scene while I was a graduate student at Harvard. But I wasn't calling Chris on this day to talk science. I wanted to talk to him about art—the original engravings from Captain Cook's Tahiti expedition of 1768.

I had first met Chris in Australia when we shared an office at the University of Melbourne. I was fascinated to learn that he was botanical editor of the first color printing of the copper engravings from Cook's 1768 voyage on the *Endeavour*. Cook had been dispatched to Tahiti to record the transit of the planet Venus. It was believed that his data, when compared with similar observations to be taken in Egypt, would allow, by triangulation, an extremely accurate measurement of the circumference of Earth. This, in turn, would greatly improve the accuracy of map making and navigation. His orders, however, extended beyond astronomy:

You are also to be careful to observe the nature of the soil and products thereof, the beasts and the fowls. . . . You are to bring home specimens of each, as also such specimens as the seeds of trees, of fruits and grains as you may be able to collect.[56]

To help carry out the latter instructions, Sir Joseph Banks arrived on the scene. Young, adventurous, and rich, Banks arranged to accompany Cook's voyage as naturalist. With him he took Carl Linnaeus's prize student, the Swedish scientist Dr. Daniel Carl Solander (1733–1782), and a large amount of equipment.

What Banks returned with eclipsed any astronomical observations in scientific significance, for he and Solander had made the very first scientific collections of Tahitian plants. These plants were so unlike anything that had previously been observed by Europeans that their scientific importance as specimens can only be compared to that of the moon rocks returned to Earth by the Apollo astronauts. Each precious specimen was lovingly brought to Solander, who did a thorough botanical diagnosis. The specimen was then placed in the hands of one of the greatest botanical illustrators who ever lived, Sidney Parkinson. The botanists worked in extremely cramped quarters, which they shared with four other crew members, for three years. The _Endeavour_ was only thirty-three meters long and ten meters wide, scarcely larger than two city buses placed side by side. After seeing the _Endeavour_ upon its return to England, Samuel Johnson remarked, "Going to sea was like going to jail with the added prospect of drowning." Parkinson, however, didn't live to hear Johnson's witticism: bitten by a mosquito off the coast of Java, he contracted malaria and died at the age of twenty-five.

Though the artist was dead, his portfolio of nine hundred drawings and watercolors was carefully bound and sent to London. Elected President of the Royal Society, Sir Joseph Banks commissioned eighteen of the finest engravers in the world to inscribe Parkinson's images in copper. Banks put up £10,000 to finish the drawings and an additional £7,000 to engrave them, an extraordinary amount of money in those days. The resultant exquisite engravings represent the apogee of botanical illustration. The swirl of scientific and social activities engaging Banks, however, resulted in the plates never being published. Eventually they found their way to some dark cabinets in the British Museum, where they sat, prey to dust and corrosion from London smog, for nearly two centuries.

In 1978, Chris Humphries, together with graphic artist Nigel Frith, conceived of a project to publish the plates using techniques and tools that would have been employed in Banks's time. An eighteenth-century printing shop was constructed on the east side of London. The edition was strictly limited to one hundred images of each copper plate, and tremendous care was taken to ensure complete accuracy of the colors and continuity from print to print. The resultant engravings, over two hundred years in the making, are breathtaking in their beauty. Ninety-nine of the one hundred sets sold immediately, mostly to art museums and other institutions. But I knew that one set, purchased by James MacEwen and Associates of London, an art dealership, was being split up and sold as singles.

I finally got through to Chris at the Natural History Museum in London. "I was thinking how Captain Cook returned with many artifacts from the Polynesians, but left them precious little save a few social diseases. What do you say we even the score?" I asked.

"What do you have in mind?" Chris replied.

"I'm trying to raise money to save a thirty-thousand-acre rain forest in Samoa. Do you think James MacEwen and Associates might be willing to donate 10 percent of their proceeds from the engravings to the project?"

Chris replied cheerily. "You know, Paul, James might just go for that sort of thing. He is heavily involved in conservation and is on the board of the Flora and Fauna Society of Britain. Let me see what he says."

A few days later, a call came from London. A very distinguished voice spoke. It was James MacEwen.

"Chris Humphries told me about your proposal for the Samoan rain forest. I think it's grand. We'd love to help," James said graciously. "And I think Chris might be willing to give a lecture on the plates to your university."

An exhibition of the Society Island Plates was held at BYU. Chris Humphries flew over to give the lecture for the opening gala. A catalogue showing the prices of the prints (between $1,400 and $3,000 each) was discreetly given to individuals expressing an interest in purchasing them. Very few prints were sold during the exhibition, with Ken Murdock and his friends purchasing those that were. After the exhibition closed, however, people called from as far away as California to make purchases totalling $90,000. James MacEwen and Associates thus presented to the BYU rain forest fund a check for $9,000.

I finally had in hand all the funds needed to pay for the school, and flew down to Samoa to negotiate the covenant. Before I left Samoa to raise the money, the village leaders and I had sketched out what such an agreement might look like—prohibition of logging, protection of the plants and animals, and a guarantee of village sovereignty over their own lands. But now was the time to flesh out the details of these broad understandings. Arriving in Falealupo, I was pleased to see that the chiefs had been absolutely faithful in their protection of the forest. Once Fuiono had run to the logging area to stop the bulldozers, not a single additional tree had been cut. And the ban on the killing of flying foxes had, from all accounts, been honored as well.

The village chiefs assembled at night in the Avatā part of the village, in the meeting house of Solia Papu Va'ai, the son of a former prime minister. Every chief in Falealupo was there, numbering thirty in total, arranged on mats around three sides of the hut. On the remaining side I sat alone. It was only then that I realized the obvious imbalance in my negotiating stance. How could one person hold his own against thirty chiefs who have been trained from birth in the art of argument and rhetoric?

Solia's house was rectangular, without walls, but with a concrete floor that can cripple you if you sit cross-legged on it for long. As I entered, a long mat was brought for me to sit on—a few millimeters of dry *Pandanus* cushioning.

The mood in the room was as hard as the concrete. As I glanced around the hut, I noticed a few new faces, some of whom had obviously been invited just for the negotiations. High Chief A'eau Tau-ulupo'u, slender and well-spoken in both English and Samoan, was a CPA and former Speaker of the House in the Western Samoan Parliament. High Chief Solia Papu Va'ai owned the Vaisala Hotel near Asau. Even though he didn't live in the village, his title was from Falealupo, and the people trusted him greatly.

Fuiono Senio began the meeting after a short speech of welcome. I showed him and A'eau the certified BYU check made out to the Development Bank for the school mortgage, and the check to the sawmill for the village debt. They both smiled. I then read the draft covenant that I had written in Samoan.

The Falealupo Covenant

We, the chiefs and orators of Falealupo, Savaii as the recognized authorities and leaders of Falealupo village, hereby affirm that we are legally and culturally empowered to represent Falealupo village in entering into a covenant with Mr.

Rex Maughan, Mr. Ken Murdock, Dr. Paul Alan Cox, and other interested donors for the purpose of preserving forever the rain forests of the Falealupo peninsula.

Responsibility of the Donors

1. In consideration of the importance of the unique beauty and nature of the Falealupo rain forest, we—Mr. Maughan, Mr. Murdock, Dr. Cox, and the other donors—covenant to assume the current debt for the construction of Falealupo Primary School as carried on the books of the Development Bank of Western Samoa and the accounts of Samoa Timber Products.

2. We, the donors, also covenant to attempt to raise further funds to form a perpetual endowment to assist in the development and preservation of the Falealupo rain forest.

3. We, the donors hereby affirm the perpetual sovereignty of Falealupo village over the Falealupo rain forest and renounce any claim or title by ourselves or by our heirs to the rain forests of Falealupo village.

Responsibility of Falealupo Village

1. In consideration of the funds and goodwill freely given by the donors, we, the chiefs and orators of Falealupo, covenant and promise to preserve forever in its pristine state the rain forests of Falealupo.

2. We, the chiefs and orators, further promise to preserve and protect the indigenous flora and fauna of the rain forests and specifically promise to prohibit the destruction and hunting of the Samoan flying fox *Pteropus samoensis* and the white-necked flying fox *Pteropus tonganus*.

3. We, the chiefs and orators of Falealupo, also promise to allow the Government of Western Samoa to formally recognize this covenant by decree or Act of Parliament and further covenant to allow the Government of Western Samoa to declare our rain forests as a National Park, subject only to continuing recognition of our ownership and sovereignty over these lands.

4. We, the chiefs and orators of Falealupo, covenant to allow in perpetuity Dr. Paul Alan Cox and his associates access to our rain forests for the purposes of scientific research including the limited and nondestructive harvesting of scientific and research specimens.

Understandings

1. The chiefs and donors agree that limited cultural uses of the forests, including collection of medicinal plants, selective harvesting of trees for kava bowls, canoes, and house construction, may continue as long as *(a)* traditional techniques and tools are used, *(b)* the uses are limited and do not significantly alter the pristine character of the rain forests, and *(c)* no chain saws, bulldozers, or other devices driven by internal combustion or electrical engines are used. The donors and chiefs further agree to allow traditional garden plots to be used along the edges of the disturbed forest as long as these gardens are for subsistence use and do not involve the clearing of primary forest.

2. The chiefs and donors agree that indigenous flora and fauna will be otherwise protected against harvesting and hunting, although fishing and the hunting of feral pigs and other noxious nonindigenous animals will be allowed if such activities are designed to protect the forests.

3. The chiefs and donors agree that all terms of this covenant shall be binding from the date of signature upon them and their heirs. They further agree that the signatures of the assigned representatives shall be considered binding upon all donors and chiefs regardless of their signature or physical presence at the ceremonies.

4. The chiefs and donors agree that the donors may use other Universities, Foundations, Banks, or other entities as conduits for their donations to meet their responsibilities and hereby acknowledge with thanks the kindness of Brigham Young University for its good offices in these regards.

As I read, many of the chiefs listened with their eyes closed: Samoan chiefs, perhaps because of their experience with kava speeches, have an amazing ability to memorize long passages of written text after only a single hearing. After I finished reading, A'eau smiled. "Very good," he said. "Now please read it again slowly so that we can go through it point by point."

When I got to the word "forever" in the first paragraph A'eau stopped me. "There," he said, "I have a problem with that word."

"Why?"

"Only God is forever. Man is limited in time."

"Yes, but the intent of a national park is to preserve it forever. All of the American national parks are preserved forever."

"This isn't America," Fuiono Senio sternly interjected. "This is Samoa. We can't guarantee what will happen in the distant future. We can only guarantee what will happen in our own lifetimes."

My mind reeled. All of the national parks I had heard of had been established in perpetuity. None had time limits on them. But the chiefs seemed insistent, and their argument against preserving the park "forever" made me realize their seriousness about strictly obeying the terms of the covenant.

"O.K.," I responded. "How about a hundred years? Will you agree to preserve the rain forest for a hundred years?"

"There isn't a person in this room who will be alive a hundred years from now," Fuiono Senio said aggressively. "In Samoa, we can't control from the grave what future generations will do. Can you do any better?"

I immediately scooted forward on the mat, the way I had been taught to respond when challenged by a chief, having been trained to avoid any sign of weakness in a kava ceremony. That maneuver apparently didn't apply to land negotiations: instead of apologizing or backing off, as would have happened during kava, Fuiono Senio moved several feet toward me, saying, "If you want to move close to me, I'm happy to move close to you." He spoke in colloquial Samoan as if challenging me to a fight, staring me in the eyes. After a minute of tense silence, I looked down and fingered a piece of *Pandanus* on the mat, while Fuiono continued to glare at me. The *fale* seemed filled with distrust, and I feared that the negotiations had ended.

Finally, Solia, who initially seemed to be dismissive of my offer, broke the impasse. "How about twenty years? We'll protect the forest for twenty years."

"I don't think the donors will go for that," I responded. "This is a lot of money we're talking about. How about fifty years?"

"Thirty," Solia countered.

"O.K. Let's go back to a hundred years then," I responded.

"Fifty years," Fuiono Senio slowly said, glaring less than a meter from my face. "You've got your fifty years."

I continued reading the covenant, penciling in changes where the village chiefs identified sensitive issues. Interestingly enough, one issue of concern was the designation of the Falealupo forest as a national park. "Why not allow Parliament to designate the preserve as a national park?" I asked Fuiono Senio. "It would greatly increase tourism potential."

"We don't want the government to have anything to do with our land," Fuiono said.

The other chiefs were equally insistent. They wanted any reference to the government removed from the covenant. I deleted the entire paragraph.

When I read the section about protecting all flying foxes, they smiled and readily assented.

When we reached the paragraph on permission to continue to study Samoan medicinal plants, I told the village that I intended to share with them any proceeds from patents or from any other breakthroughs in discovering new drugs.

"How much?" A'eau asked.

"What do you think is fair?" I responded.

We agreed to a village share of 30 percent of royalty income should one of the medicinal plants I collected in Falealupo ever become commercially marketable as a drug. This part of the covenant was of special importance to me, as I wanted to demonstrate to the National Cancer Institute and our joint patent lawyers that my work was encumbered by obligations to the village. This would increase my ability to obligate the U.S. Government to share patent royalties with the village, something that had never been done before.

We came to the section on hunting. Could the village continue to hunt fruit pigeons in season? We would have to refer that to the other donors, I said, as many foreign conservationists are birdwatchers.

By midnight, after some hours of highly detailed and arduous negotiations, we finally had a full agreement. The Falealupo forest would be protected for fifty years, but as a preserve controlled by the village, not as a national park. If all went well at the signing scheduled for the following week, the Falealupo Rain Forest Preserve would be established.

A

10 cm.

C

B

8 mm.

2
mm.

4
mm.

E *D*

Psychotria insularum

CHAPTER 11

Metamorphosis

*A crowd of small metamorphoses
accumulate in me without my
noticing it, and then, one fine day,
a veritable revolution takes place.*

Jean-Paul Sartre, *Nausea*

The next day the film crew arrived: a husband-and-wife team from Australia, Paul Tait and Jeni Kendell, an Australian soundman, and a Swedish director, Marianne Landin. The village had previously given permission for the crew to produce a film in Falealupo about conservation and ethnobotany, but I was still nervous about how they would fit in.

I first met Marianne and her husband, the famous Swedish filmmaker Bo Landin, at an Uppsala dinner party. Bo had won every major European environmental film award and had an international reputation as an environmental advocate. Representing, as he did, a consortium of Swedish TV, Channel 4 England, and the Australian Broadcasting Corporation, Bo assured me that his group would be happy to retain the original Samoan of the speakers, since most of the films and TV programs in Sweden used subtitles anyway. Bo was also interested in filming the use of *matalafi (Psychotria insularum)*, a plant that Lars Bohlin of Uppsala University and I had found to be pharmacologically active.

Since Bo himself had a previous commitment to film in the Antarctic, his wife Marianne agreed to come along with the Australian filmmakers.

"Sounds great," I said with some trepidation.

In most respects, the film crew adapted extremely well to village life. But there was one aspect of Samoan culture that I forgot to warn them about. Paul Tait asked that Sunday afternoon if it would be O.K. if the film crew went for a swim.

"The village frowns on people swimming here on Sunday, but I don't think anyone would mind if you hiked through the forest to Fagalele beach, about a mile along the coast. No one lives there, and it really is quite pretty."

The film crew made the journey, but unbeknownst to them and to me, Pela sent two or three children to quietly trail them, just to make sure that they didn't run into any trouble.

The children returned first. From the looks on their faces I could see that something was seriously wrong. They wouldn't speak to me, but went straight to Fa'asaina in the cook hut. Fa'asaina came out looking alarmed and went to Lamositele. Like some sort of unseen contagion, a look of concern spread over his face, and he went and spoke to the children.

"What's up?" I asked after the children had left.

Lamositele looked at me with a resigned expression, and then said, "Your friends. They're swimming nude at Fagalele beach."

"Good grief!" I said. "I'm sure they didn't realize they were being observed."

"It was quite a shock to our children, who had never seen naked *palagi* people before," Lamositele replied.

Contrary to European myth, the Samoans, like most Polynesians, are very modest people, but their traditional sense of modesty is different from that of Europeans. In old Samoa, women commonly went bare-breasted, and even today in the remote villages women still do their laundry in the rivers and bathe without their blouses. But the area from the navel to the knee is always kept covered.

Paul and Jeni were horrified when I explained the breach of Samoan etiquette to them. They hadn't known that anyone had followed them. "The rain forest not only has ears, but eyes," I said, remembering Bill and Dixie's experience on Mount Fuionō. Samoans have a saying: *"Ua iloa se,"* "Even the crickets know."

We spent several days filming, but I was preoccupied with preparations for the visit of the principal overseas donors. Ken and Rex would come for sure, and King Gustaf and Queen Silvia had even toyed with the idea of traveling to Samoa incognito. Weighing heavily on my mind was the need to serve as orator for the visiting party during the upcoming kava ceremony. Although I had spoken at kava ceremonies before, I realized that my remarks at this particular one were crucial. Not only did my speech need to comply with the dictates of Samoan rhetorical form, but it also needed to promote village integrity in adherence to the covenant provisions. Since the donors and I were entering into the covenant with the village, it would have been inappropriate to have engaged the services of a village orator to represent us during the ceremony. And bringing an orator from another village was not appropriate, given the sensitive nature of the negotiations concerning village lands.

Two days later, I left for Pago Pago to meet the donors who had come for the signing. Returning to Savaii, we filled the small Norman-Britten Islander aircraft chartered by Rex Maughan. I asked the pilot to overfly the Falealupo rain forest so Rex, Ken, and the others could see it. There was a gasp in the plane as the lush green forest came into view beyond the jagged black interface between lava and reef. From the air, the large rounded crowns of the banyan trees looked like giant cauliflowers emerging above the forest canopy. Other canopy-level trees, such as the tall *maota* tree, were less rounded and full, consisting of a few large twisted branches. Now and then we could see small natural gaps in the forest filled with banana-like *Heliconia paka,* seeded bananas *(Musa acuminata),* and young saplings just beginning to grow. We could also see white long-tailed tropic birds lazily floating in pairs high in the sky. The aircraft bumped with turbulence, and then the scar of the logging came into view.

"That is what we are fighting," I said.

The plane looped back low along the coast, and the pilot waggled his wings in response to the waved greetings from the villagers. The landing on the little dirt strip in Asau was rough. We could barely fit all of the passengers and their luggage into my four-wheel-drive van, but somehow we made it. We turned onto the Falealupo road in the cool of the early evening and drove down the single-lane road, arriving at the rain forest just as it was getting dark. I parked the van and opened the windows so the cool air and the sound of birds could fill the vehicle.

The village chiefs' council had arranged for each of the visitors to be placed with a different family representing a different part of Falealupo. Ken and Kristin Murdock stayed with Mao, the mayor, in Malaetele, while Rex Maughan and his group were hosted by Soifuā in Faleolo. Jim Winegar, Dan Wakefield, and their friend Phil Goodrich stayed with Solia's family in Avatā. The film crew was happily ensconced with Seumanutafa's family in Vaotupua. Thomas Elmqvist and several American reporters completed the entourage. All of the guest houses were gaily decorated with flowers and vines, and the sounds of laughter and singing filled the village as the sky grew dark.

After dinner, I drove down the sandy road through the village. Falealupo that night was bathed in magic. The stars were brilliant in a Van Gogh sky, far from any pollution or artificial light source that would have obscured their radiance. The white beach sand was iridescent, reflecting the first light of the moon. The thatched *fale* appeared like luminescent mushrooms, glowing with the light of Coleman lanterns.

After seeing that all of the visitors were taken care of, I walked down the beach to a quiet spot and gazed out at the dark surface of the sea. As the gentle waves washed my feet, I thought of the kava ceremony on the morrow, and rehearsed my speech one last time.

Pela approached me as I drew water from the concrete tank for my shower. "Koki—be careful tomorrow with the chiefs' council."

"Why? Are they unhappy with the covenant?"

"They may try to entrap you. Don't accept a chief's title from them—let our family give you one. Be careful of getting too wrapped up in village affairs."

"I think they will present titles to Ken Murdock and Rex Maughan, but I'm safe. After all, I'm officially an untitled man of Falealupo and have a mat to prove it."

The day of the covenant signing began with a festive air. At the first rays of dawn, the villagers were up and busy, gathering torch ginger and *Heliconia* flowers to weave into the coconut leaf plaitings on the meeting houses. The sweet smoke of pigs and taro in the *umu* filled the air. The village carpenters had finished the new school in record time, with work having commenced only a few months earlier. The village had asked that the day begin with an inspection of the school by the donors. With chiefs on either side carrying kava roots, we walked down the road together.

As we approached the school, on the border between Avatā and Malaetele, I was struck by the tremendous effort that the villagers had made to prepare it for exhibition. The white sand in front of the school, comprising an area about the size of a football field, had been swept completely clean. The school was decorated with hundreds of hibiscus flowers, threaded on the pinnae of coconut leaves. The principal, a quiet, self-effacing man, strolled out to greet us. He was wearing a carefully pressed and starched shirt and an elegant lavalava made out of black suit cloth.

"We are pleased to welcome you to the Falealupo Primary School," he said in English.

Several women placed flower leis around our necks and invited us into a classroom. As soon as we entered, the students burst into song. Several schoolchildren placed heavily scented *moso'oi* leis around our necks. Rex and Ken were obviously pleased, and Ken's wife, Kristin, looked completely enchanted. As I listened to the children's songs, I recognized the original compositions of the male teacher who was leading the choir. There is nothing quite like a traditional Samoan chorister turning conducting into playful dance, like Duke Ellington whirling and spinning in white sequined tails in front of his big band orchestra. The Samoan conductor was stripped to the waist, with his body tattoo just peeping above his flowered lavalava. In contrast to the polite and cautious motions of the uniformed schoolchildren who sat around him on the ground, his motions were energetic and exaggerated. He made a fist and a hitting motion to indicate the rhythmic emphasis, and seductively wiggled his hips during a particular lyrical passage.

Within a few moments, several villagers leapt up and began dancing to the music. Rex and Kristin joined the fray, and soon nearly everyone present was swirling and dancing to the children's music. The two different film crews, one making the film documentary and the other a news crew recruited by Rex, were trapped amid the gyrating bodies on the flower-filled dance floor, but they too soon joined in, cameras pressed to their faces. After the dance, I made a brief speech of appreciation to the faculty and schoolchildren, and we continued on our way.

We were directed to a circular *fale* in Malaetele that had been decorated with flowers and woven coconut trim. The village chiefs were assembled in a different *fale* about two hundred meters away. After sitting on the mats, we received a message from the chiefs: they wanted to see me first alone. I walked across the sand to the assembled chiefs, entered the *fale,* and listened to the traditional greeting rhetoric. After I answered, the chiefs smiled, and Fuiono Senio began to speak.

"Are the donors happy?" he inquired.

"Yes. They had a lovely night and are thrilled with your hospitality. The village is very beautiful."

"Good," said Fuiono. "Are they also satisfied with the covenant?"

"Yes. They went carefully over the revised document and found it acceptable. I understand that the village will be conferring chief's titles on Ken Murdock and Rex Maughan."

"We will be conferring three titles," Fuiono said.

"Three?" I said. "Who is the third one for?"

Fuiono paused and then said very firmly. "You. We want you to become Nafanua."

I panicked. Accepting a chiefly title, particularly a high and legendary title like that of Nafanua, would violate all of the dictates I had learned as a scientist in dealing with indigenous people. As an ethnobotanist, I came to Samoa to learn, not to become an agent of change within the culture. By fighting the loggers, I had already stretched my scientific objectivity to the breaking point. If I accepted the Nafanua title, my relationship with Samoa would forever change. I would assume lifelong responsibility for Falealupo village in bad times and good. I would no longer be free never to return, if I wished, but would be duty bound to serve Falealupo and Samoa forever. Up to this point, any service I had rendered to Falealupo had been on my own terms; acceptance of a title would be the equivalent of entering marriage vows with the entire village. And what would my colleagues think? Already I had replaced scientific objectivity with conservation and cultural preservation as the overarching ethic of my work. I knew very well that conservation efforts didn't count for much in the academic world. My acceptance of the Nafanua title would set me up for accusations that I was manipulating the culture for my own ego, becoming a modern Lord Jim.

I replied haltingly to Fuiono, "I am deeply honored that you would wish to bestow a title on me, but I have not performed a sufficient *tautua* for the village, and . . . "

Obviously anticipating my objection, Fuiono cut me off. "You are qualified, and your work for the Falealupo Primary School and the rain forest is more than sufficient as a *tautua*."

Fuiono was forceful, but I feared both the added responsibility of the title and the liability it might incur for me in the scientific community. "I respectfully decline."

"I don't think you understand," Fuiono said sternly. "We are not *asking* you to accept the title. We are *telling* you to accept the title. If you refuse, then we refuse to sign the covenant."

"What!" I exclaimed. "You won't sign the covenant if I don't accept the title?"

Fuiono nodded affirmatively. I looked at the assembled chiefs and saw that they were resolute in their decision.

"But the donors have traveled so far, and are waiting in the *fale* over there. How could you imperil the covenant that we have worked so long at negotiating? And what about the rain forest?"

Fuiono did not waver. "If you don't accept the title, we won't sign the covenant, and we will not create the rain forest preserve."

I closed my eyes. The chiefs were playing hardball with me. The village believed that my foreignness, my access to power, and my attitude toward forest conservation all bespoke some connection to the ancient goddess. But instead of feeling venerated, I felt as if the chiefs were trying to entrap me, just as Pela had warned.

I had initially come to Falealupo with no intent beyond studying herbal medicine and perhaps having a long shot at finding a new treatment for breast cancer. I did not come to discover a novel AIDS therapy. I did not come to become embroiled in a village dispute with a logging company. I did not come to fight my own government for an endangered species. And I certainly did not come to interfere with an indigenous culture. And yet the title Falealupo wanted to confer on me was roughly equivalent to that of Vishnu in the Hindu pantheon. Just the thing to tell my colleagues at our next professional meeting: "Hi, I'm Nafanua, the Samoan goddess of war."

If I accepted the Nafanua title, if the Western Samoan people even knew that the district was conferring the title, my relationship with Samoa, my scientific career, and my life would be forever changed in a multitude of unpredictable ways, most negative. Nothing good could come of this.

I looked at the chiefs. They weren't smiling. Fuiono Senio in particular looked very stern. He wasn't bluffing—if I refused the Nafanua title, he would shut down the entire ceremony, just the way he had shut down the logging operation. What would he do, invite the loggers back? Yet hadn't I pledged to stop the logging, whatever the personal cost? From that decision had flowed all of these consequences. How could I turn back now?

I stared into Fuiono's eyes. "I pray that God will enable me to carry the burden placed upon me," I said as I began the ritual recitation of acceptance. With those words, my life changed forever.

As a group, we walked to the *fale* where the donors were waiting impatiently. "What happened over there?" Ken whispered to me as I sat down beside him. "Is it trouble?"

"There have been a few changes in the program, but the covenant remains intact," I whispered.

"Good," Ken replied.

A large wooden kava bowl was brought in from the side of the *fale* by the untitled men. The seventy-year-old village virgin, Soimavī, took her place behind it. A mat was dragged in front of the village chiefs, and each laid a long kava stick on the mat. The mat was then brought to the visitors' side of the *fale*. Custom dictates that the orator who will give the kava speech not say the *folafola 'ava,* so Lilo stripped off his shirt and slid forward on the mat. He paused for a moment, and then began speaking in a crystal-clear, ringing voice:

'Ua mamalu le aso ae ua pa'ia
fo'i lo tatou taeao. Ua liligo
le fogatia ae pa'ū le Tuaau
mafuamalu. Ua paū mauga ae
liligo vanu e pei o le fetalaiga
i le alofi o le Tu'imanu'a.

Our day is dignified
and our occasion sanctified.
The pigeon hunting grounds are
silent and the forest echoes with
sacredness. The mountains echo
and the valleys are sanctified as
the kava circle of the Manu'an
King.

Ae ou te manatu o lupe nei sa
vao 'ese'ese ae o lenei ua fuifui
fa'atasi. Ua moni tala a le tusi.
O le faamoemoe ua taunu'u
o le la'au o le soifua lea. Ua
o'omia nei a'u i le vaigalu,
ae fa'agāfua ia lo'u nofoaga.
Fa'amagalo lo'u leo, ae se'i ou
alagaina le galu auā e faigata
ava.

We pigeons from distant forests
are now of the same flock. The
words of the book are true—
we here behold the tree of life.
I sit between two orators who
grant me authority to speak.
Forgive me as I speak in their
presence of the sacredness of
kava.

Fai mai o le 'ava o le
motugauso ma le ata na toto.
Le ava sa fa'alafi. Ave i fale,
si'i i fafo. O le 'ava sa toto ae
fa'aaga'aga e sua auā Pule ma
Tumua.

It is said kava springs from the
union of an island with its red
reflection. This kava was long
hidden, but is now removed
from its repository, and bleeds
red in honor of the chiefs of
Samoa.

O le 'ava fo'i sa fa'atofala'i i
maota ae momoe i laoa, auā
le galuega a le Atua.

This is the kava that sleeps in
chiefs' houses and reposes in
orators' abodes in the service of
God.

O lea ua ou taulimaina le ava
fa'atupu. Le ava na afio mai ai.
Le 'ava ua uma ona teuteuina.
O le ava fa'alālelei ua mago
fa'alā. Ua i ai le Lupesina. A
ao o le malae, ua tasi ae afe.

I speak of kingly kava brought,
stored, dried, and made
beautiful. We have here special
kava—called the white
pigeon—and though it be a
single root to us it represents
thousands.

Ma o le ā fai loa le pule a le
Tapuagia auā ua lava ma
totoe. . . .

I will now exercise an orator's
privilege and distribute the
kava because we have more
than enough. . . .

As Lilo ended his speech, the kava roots were ceremoniously presented to Ken, Rex, and me. Chopped kava was then brought to Soimavī, who began steeping it in the kava bowl. After ceremoniously swirling the fiber strainer, she flipped it over her back. The strainer was caught by one of the young men, who stepped away from the hut to wring it out. After a moment or two, Soimavī put her hand over her shoulder and, miraculously, the strainer reappeared, having been expertly tossed by one of the young men. As she continued to work, and as the brew grew progressively stronger, the village's kava speeches began.

Tai'i', the senior orator, explained that two important events would occur this day. First, the village would confer three chief's titles on the visiting party. And then the visiting party and the chiefs would sign the covenant, establishing the Falealupo Rain Forest Preserve for fifty years. Becoming chiefs is such a tremendous honor, we were told, that we should no longer use our common names in Samoa, but only our new titles.

After the speeches, the kava was ceremoniously passed to Rex Maughan, who accepted the title Tilafaigā, and Ken Murdock, who accepted the title Taemā, the names of the two Siamese twins who swam from Fiji to Falealupo. And then the cup was refilled and presented to me.

The distributer of the kava, La'ulu, yelled in a booming voice: _"Taumafa lau ava, Nafanua!"_ ("Partake of your kava, Nafanua!")

I dribbled a few drops on the mat and then intoned, _"Ia manuia la'u nofo ma le vao matua!"_ ("May my reign and the rain forest be blessed!") Everyone in the _fale_ responded, _"Ia Soifua!"_ ("Live Long!"). I raised the cup to my lips, and with that single drink, I became Nafanua.

After the chief investiture was completed, we turned to the covenant. Speeches were again given, and I responded on behalf of the donors. The rain forest was considered sacred in Samoa because it was created by the hand of God, I said, and so, by protecting the rain forest, we were in effect engaging in a collective act of worship. Since we had sealed the covenant with sacred kava, it would be remembered forever.

After the applause, I read the covenant one last time. The village chiefs instructed us to sign with our new chiefly titles. Each chief then signed. We ran out of signature space on the front, so the back was used as well. With the signing finished, there was a loud scream of "Choo-Hoo!" from outside. Bedecked with coconut leaf anklets and bracelets and hung with flower leis, untitled men ran into the hut with a giant wooden bowl of hot _taufolo_, breadfruit cooked in coconut cream. The delicacy was ladled onto banana leaves and distributed. Women of the village then appeared singing and dancing around the fale, carrying large plates of food. Hot taro, chickens, pigs, lobsters, reef fish, corned beef, and *palusami* were quickly consumed by everyone and followed by Samoan cocoa.

Ritual presentation of gifts, called *sua,* to the donors began: a single coconut for each with a small amount of paper money pinned to it, followed by a fine mat and a baked pig. The first pig brought out was so large that two men strained to carry it on a stout stick thrust between its legs. "What am I supposed to do with that?" Ken asked.

"Now that you are a chief you'll be confronting a lot of decisions like this," I responded.

"Do we have to take it?" he said, looking at the gigantic carcass in front of him, its ears and mouth bedecked with hibiscus flowers.

"On that topic, you know the culture as well as I do, Ken," I answered. "To refuse a gift is to risk insult."

"But just look at it!" Ken said in desperation. "It's gigantic. It won't even fit on the plane!"

"You'll just have to take it as your carryon."

Ken gave me a pained expression. I sought a culturally acceptable way out of his dilemma.

"Chiefs and orators of Falealupo village," I began, "Taemā and Tilafaigā are deeply touched by the respect you have shown them. As a small token of their regard, they wish to give the village a gift."

"Please carry this pig to the chiefs and orators of Falealupo village," I asked the untitled men.

Fuiono smiled.

After the ceremony, we shook hands with all of the villagers, and loaded the donors into the van to drive them back to the airstrip. Where the road intersected the Falealupo rain forest, we stopped. Everyone got out and looked at the forest, which, for the first time in years, was free from peril. Jim Winegar led us in singing "Happy Birthday" to the forest, and Rex Maughan offered a prayer.

When I returned to our *fale* after saying good-bye to the donors, I was met by Lamositele and Lilo Manuele.

"Your kava speech was incredible," Lilo gushed. "There is only one error you committed."

I tried to remember the details of the day, terrified that I might have unwittingly rendered offense in the precisely articulated Samoan kava ceremony.

"What did I do wrong?" I asked.

"Where do you think that pig came from?" he asked.

"I assumed it came from the village," I responded.

"The pig," Lamositele said, "came from me. I donated it so that it could be presented to you."

"Yes," I said, not quite comprehending, "but what did I do wrong? Were the chiefs unhappy that I gave it back?"

"Oh, they were delighted to share that pig."

"Yes," I said, still trying to comprehend, "and didn't they give our family a share?"

"The pig," Lilo Manuele interjected, as a professor might lecture a particularly slow student, "was presented to you by the village in accordance with Samoan custom."

"So, what should I have done differently?"

"Next time you are given a pig," Lamositele concluded, "don't give it away to the village. Bring it home to us."

10 cm.

4.5 cm.

A

B

C

Callophyllum inophyllum

CHAPTER 12

Victory

*Nonetheless, he knew that the tale
he had to tell could not be one of a
final victory. It could be only the
record of what had to be done, and
what assuredly would have to be
done again in the never ending
fight against terror and its relentless
onslaughts, despite the personal
afflictions, by all who, while unable
to be saints but refusing to bow
down to pestilence, strive their
utmost to be healers.*

Albert Camus, *The Plague*

Bill, Thomas, and Dixie sat across from me at the table. We ordered another round of Perrier and enjoyed the Swiss sun. Verne Read had generously funded my trip to Lausanne to lobby CITES—the 1989 Convention on International Trade in Endangered Species—on behalf of Pacific flying foxes. I had flown in from New Zealand, where Barbara and I and the children had been living for a few months while I edited a book on Polynesian ethnobotany. Each of us had infiltrated separate delegations: Thomas was a special scientific advisor to the government of Sweden, I represented the World Wildlife Fund, and Bill and Dixie were delegates from the International Union for the Conservation of Nature—IUCN. We had plastered Lausanne with posters of Samoan flying foxes, held a reception for CITES delegates at which we screened Bo Landin's footage of soaring flying foxes, secured the support of all of the environmental organizations for our proposal, and then lobbied delegates individually on its behalf. We previously had submitted our proposal to the official United States representative at CITES: the U.S. Fish and Wildlife Service. True to form, the Service was critical of our proposal. What the Service did not know, however, was that we

had submitted our proposal to another CITES signatory nation—
Sweden. The Swedish delegation invited the U.S. delegation to a
strategy meeting, but the Fish and Wildlife official became belliger-
ent, insisting that the proposal be tabled until more data could be
collected. I pulled him aside and spoke quietly with him.

"You need to do what you think best, but please remember that
Sweden has equal footing with the United States at this meeting and
will go forward with the flying fox proposal despite your opposi-
tion. We have tallied votes and I assure you that the Unites States
will be the only country in the world to vote against this proposal.
And to explain why such a vote is not in the interests of the United
States, I wonder if you would be willing to speak with Senator
Hatch's office on the phone?"

The official grew red in the face. "O.K.," he stammered, "We
will support the proposal."

As Bill, Dixie, Thomas, and I toasted each other with Perrier, we
savored our victory. All of the signatory nations to CITES unanimously voted to stop commercial traffic in *Pteropus samoensis* and
six other species of Pacific Island flying foxes. They also voted to
monitor closely the trade in numerous other species. During the last
twelve months, our efforts had also been rewarded with the creation of a U.S. national park, and only six months previously, the
Falealupo Rain Forest Preserve had been created by covenant with
the village.

"Amazing what you can accomplish with smoke and mirrors,"
Bill said. We all laughed.

When I returned to New Zealand, Barbara, the children, and I
packed for another six months stay in Samoa. This would be the
last year of my Presidential Young Investigator Award. At the Auckland airport, the Air New Zealand ticket agent struggled to lift our
bags. "What do you have in there?" she asked.

"Schoolbooks for our children," I replied. "We'll be living in a
remote village in Samoa working on rain forest conservation, but
we need some way for our children to keep up in school. I'll be
pleased to pay an excess baggage charge."

The agent softened. "That won't be necessary, Mr. Cox. Have a
good flight."

We spent the night at the Bethams' and the next morning drove
to the Savaii ferry. A new vehicular ferry, the *Lady Samoa*, had
recently been sent to Samoa with Australian aid funds. It was spa-
cious, and a tremendous improvement over the previous service to
Savaii. Also improved were the new advance reservation and book-

ing systems that precluded the long waits on the wharf that we had been accustomed to.

The sun glistened off the water. In the distance I could see the smaller islands of Manono and Apolima. Silhouetted against the horizon was Savaii, which always looked dark and foreboding.

The children were clearly happy to be in Samoa again, as was Barbara. Mary carried her Christmas present—an inflatable raft with a viewing port—and was anxious to use it on the Falealupo reef. But for some reason I had grown anxious that something might happen to the children, and was wary about them venturing out in the ocean.

Our January schedule was full. First, some friends from Utah would come for a short vacation. Two weeks later, a group of Thomas Elmqvist's students from Sweden would visit so that I could give them an intensive course in tropical ecology. Later in the month, Rex Maughan, Jim Winegar, Dan Wakefield, and others of the original Falealupo donors would be coming down to celebrate the first anniversary of the Falealupo covenant. After the donors departed, I would again be free to resume my medicinal plant research.

Lee and Kristy Phillips and their children seemed to have a wonderful time the next week and became good friends with some of the villagers. As they got ready to return to America, I suggested that their fifteen-year-old son Revell take home an unusual souvenir: a full-body tattoo. The villagers thought it would be great fun. Kristy, hoping that I was joking, expressed some concern. But I told her that what the village had in mind would be a real Samoan tattoo from breast to knee with only one minor modification—it would be drawn with a permanent marking pen rather than with tattoo implements. The family led Revell to the small men's house, where he was stripped to the waist. The _tufuga,_ or tattoo artist, smiled as I handed him one of my permanent black laboratory markers. He began drawing with a slow and steady hand on Revell's exposed flesh. "Too bad this isn't for real," the tattoo artist explained, "it would look very good on him."

"Next time," I said.

Fa'asaina and another woman came to the hut and started singing to Revell the traditional songs of courage that are used to comfort a man who is enduring a painful tattoo, except that sometimes they started giggling. Real Samoan tattoos take several weeks, the pain is intense, and there is great temptation to seek respite by quitting. But stopping a tattoo before it is finished is regarded as cowardly, and subjects the quitter and his family to ridicule with the most intense pejorative known in the Samoan language: *pe'a mutu* ("tattoo coward"). The resultant derision can destroy the social standing of the entire extended family. Several of my Samoan friends have told me that although they began their tattoos out of personal vanity, they completed them only so that their families would not suffer ostracism.

Revell's pseudo-tattoo was finished within an hour, just in time for Lee and Kristy to make the afternoon ferry to Upolu. Once at sea, Revell removed his shirt to gasps of amazement and then laughter from the Samoan passengers.

After the Phillips family left, we had a couple of weeks in the village without visitors from the outside. We were all looking forward to this time. In many ways, we began to feel as if we were villagers too. The children enjoyed wandering around the village, exploring the lava tubes, beach, and forest. I spent most of my time interviewing healers, but was able to take an afternoon fishing trip with Paul Matthew to Fagalele beach. And all of us enjoyed swimming in the tidal pool nearby. In some respects, the children were even more adventurous than I. When Pela's grandson Lui brought some live coconut larvae to the *fale,* all of the children ate a few. They played board games every afternoon with the village children and seemed to make close friends. Barbara drew very close to Fa'asaina and enjoyed visiting with the other women in the village. Since we had now made several extended trips to Samoa as a family, our novelty as resident foreigners had subsided. We had begun to fit, at least in some degree, into the pace of village life, although, unlike the other villagers, we could come and go at will, choosing village life or our previous more Western existence.

Only two other families in all of Falealupo had a car. High chief Seumanutafa Nu'umau, a neighbor of Lilo Manuele, owned one— a rusted pickup truck to which someone had bolted a pair of blood-tipped cow horns. Walking along the little dirt road in the forest, I would sometimes see Seumanutafa Nu'umau's adult son Asovao driving the truck down the road, with the back filled with ten to fifteen people, some hanging on over the edge as if in a college stunt. Yet this was no feat to be bragged about to one's roommates: that little truck represented one of the few conveyances that the villagers could use to travel from the main road to the village without walking the five long miles.

For local transportation, the only alternative to Seumanutafa Nu'umau's truck was Togia's van. Togia was half Chinese and half Samoan. Although he did not play a significant role in the local power structure, his carefully painted European-style house was one of the largest in the village. Attached to his house was the dream of nearly all aspiring Samoan entrepreneurs—a tiny store, with canned goods neatly stacked on a few shelves behind the open-air counter. But the most glorious of all of Togia's possessions—even more wonderful than his newly painted house or his store—was his bright red twelve-seater Toyota van. It was this van that was the ultimate source of most of his ongoing prosperity.

He drove his van in a sixteen-mile circuit to Falealupo village from the main road at least twice a day. Unlike Seumanutafa Nu'umau's pickup, there was no frivolity in this conveyance. Each passenger carefully took a seat, still covered in the original plastic, and paid for the privilege.

Although I had never ridden in Togia's van, I had became aware of the customs it involved when I inadvertently intruded on its niche during one of my first days in Falealupo several years before. Driving down the Falealupo road, I stopped my vehicle out of courtesy when I saw a group of older women walking down the road.

"_E tautala ta'avale atu, ia outou maliu mai ina ia ou momoli atu ai outou i lo outou fa'amoemoe._" ("With all due respect, come and I'll drive you to your destination.")

The women silently entered through the sliding door and carefully took their seats. I proceeded down the road to the seaside, where the women asked to be let off.

"Thank you very much, sir," they said respectfully in Samoan as they were leaving. Then one of them pressed a wadded-up two _tālā_ bill into my hand.

"Oh no," I protested, "I couldn't accept this. The ride is for free."

"Don't you need some money for gasoline?" she said.

"No, this vehicle runs on love," I half-joked.

Word spread that I was quite happy to give rides to anyone free of charge if they were going my way. And after a while, the villagers stopped offering me money. Togia was always polite to me, if somewhat aloof, but he may have perceived me as a threat to his business.

Material possessions weren't the only difference between our family and the village. Our access to money, medical care, and technology (such as our vehicle) put us on very unequal footing with the people of Falealupo. And despite our attempts to be sensitive to local attitudes, in many ways we remained deeply Western. I assumed that, given the opportunity, the people of Falealupo would have sought the benefits of Western life we enjoyed, but in this assumption I was mistaken.

The ability to drive a car, for example, is rare in Samoa. Perhaps because Samoans anciently celebrated voyaging (the eighteenth-century explorer Bougainville termed Samoa "The Navigator Islands" after its people's skilled seamanship), they continue to demonstrate respect for anyone who pilots a conveyance of any kind. Every so often passengers call out *"Mālō le fa'auli!"* ("Congratulations on the steering!") or *"Mālō le silasila!"* ("Congratulations on your watchfulness!"), to which the driver is expected to respond, *"Mālō le tapua'i!"* ("Congratulations on your support!"). The driver is honored not only because of his responsibility for passenger safety, but also because he pilots a craft of unbelievable expense in a nation that routinely slaps a 120 percent import duty on all vehicles. For that reason, few Samoans in remote villages know how to drive, and few ever ask for the opportunity. In fact, in all of Falealupo village, there were only five drivers: Barbara, myself, Togia, his son Viliamu, and Seumanutafa Nu'umau's son Asovao. One day I decided to add one more villager to the list: Uati Taofinu'u.

Uati was the son of one of the village's paramount orators, Taofinu'u. Like all Samoans, he used his father's title as his surname. Uati had a warm and penetrating smile—not unusual for a Samoan child—and could often be seen doing what Samoan boys do—throwing rocks to knock mangoes down from the trees, playing with wheeled vehicles constructed from sticks and the remains of corned beef cans, or carrying a long bamboo fishing pole and line to the sea. But there was something unusual about Uati. His left leg was stunted, with his foot turned about twenty degrees out of proper alignment, and his knee seemed unable to bend. I suspected that the deformity was congenital. Uati could walk only with difficulty, and I often saw him hobbling along, trying to catch up with his playmates.

Uati would wave and call out my name as I passed his small hut. Eventually I inquired among some of the other youths if they knew him.

"You know, Uati, the son of Taofinu'u," I asked.

"Oh, you mean Uati the clubfoot!" they giggled.

Although adult Samoans never mock someone who is infirm, Samoan children, like children everywhere, are conscious of those who are different and are sometimes mean-spirited.

One day I was driving up the Falealupo road when I saw Uati wave and hobble up behind the vehicle. I stopped and motioned him up to my window.

"Sir Uati," I said to him in a respectful greeting not commonly used for children, "can I offer you a ride?"

There was a titter among the village youths who were packed into the back of the van, and I heard one softly say to me, _"Ua tumu"_ ("We are full").

"Unfortunately there is no more room for passengers," I responded. "But I wondered if Sir Uati might consent to drive."

Uati looked at me in wonderment, fearing another jest.

"Uati," I commanded, "come sit up here on my lap and put your hands on the steering wheel."

I lifted Uati into position and whispered to him, "Don't worry, I'll operate the pedals."

With Uati steering, we drove up the bumpy road. His face was full of excitement. After a few minutes I let him off, and in the rear-view mirror, I could see him dancing with delight, in his staccato way, as he went home to tell his family of his adventure.

Subsequently I stopped to let Uati "drive" my car whenever I saw him, the only villager I ever accorded this privilege. As I watched Uati struggle to keep up with his peers, I began to wonder if an operation could increase his mobility.

I contacted the Shriners' hospital in Honolulu, which I knew had a policy of admitting crippled children without regard to their ability to pay. Would Uati Taofinu'u be a candidate for corrective orthopedic surgery?

The staff thought he was. All they needed was the written consent of his parents to evaluate and treat him.

I approached Uati's father, Taofinu'u, privately about the matter. Taofinu'u is very jovial and known for his acumen at chiefly repartee. But he had never expressed appreciation for my interest in his son. Perhaps he felt embarrassed by his son's condition. Explaining the Shriners' program, I told Taofinu'u that the surgeons in Honolulu believed his son might have a correctable condition, and that Barbara and I would be pleased to pay Uati's airfare there.

Taofinu'u, renowned for his wisdom in village affairs, fell silent, surprised by the offer.

"Let me think about it," he responded.

The next time I saw Taofinu'u, I inquired if he had considered my offer.

"Yes," he said. "I don't think it would be a good idea."

I tried to discover the reasons behind his refusal, but Taofinu'u, in a firm but courteous way, let me know that he did not wish to discuss the situation further. I left our exchange puzzled. Why would Taofinu'u refuse an offer of medical assistance for his son? We had found previously with Lolina, the cross-eyed five-year-old girl in Upolu that our doctor friend John had refused to see, that Samoan villagers are sometimes frightened about surgical procedures. After

Dr. John had left Samoa, Barbara and I found a qualified ophthalmologist at the L.B.J. Hospital in Pago Pago, American Samoa, who agreed to see Lolina. All we needed to do was to fly Lolina and her mother over from Upolu. When we offered them air tickets for the visit, Lolina's parents turned us down flat. After considerable probing, we discovered that they feared that the ophthalmologist, upon seeing Lolina, would *sale ona mata*—"scoop out her eyes." Better their daughter have crossed eyes than none at all, her parents reasoned. When I finally learned of these objections, I reassured them that the ophthalmologist would not remove Lolina's eyes, nor would he perform any medical procedure without their express consent. Finally we were able to coax Lolina's mother to take her to Pago Pago, where the doctor prescribed corrective lenses.

Perhaps it was out of a deep sense of pride that Taofinu'u refused a handout, even for such an important purpose. But I wondered if his objection centered on the corrective attempt itself. It was as if Taofinu'u had long ago acquiesced to his son's condition, an acquiescence that brooked no reversal. Westerners are often struck by the deep, often religious, fatalism of many non-Western peoples. In many indigenous cultures, one learns to bear changes of fortune quietly and without protest.

Resignation is not, of course, synonymous with acceptance. But stoicism is instilled in Samoans at a tender age. Even little children are required to stop weeping immediately after an injury. Rare is the Samoan who will emit a cry of pain, even after grievous injury. When one of the film crew slammed Lilo Manuele's hand in the door of a vehicle, he quietly tapped them on the shoulder, pointed out his situation, and calmly waited for them to open the door.

Thus I do not consider Taofinu'u's stance bereft of dignity. At some time each of us must confront similar choices in the face of adversity: resignation or hope for an unseen redemption. Few consistently accept loss or unfailingly pursue victory, as constant resignation leads to fatalism, while consistent but unrequited hope provokes either cynicism or an overly optimistic view of the world, like that of Dr. Pangloss in Voltaire's *Candide*. Disbelief breeds in both the unchurched and the overly zealous. Perhaps only those meek enough to seek transcendence but bold enough to accept reversal can ever truly enter the realm of faith.

The Falealupo forest was extraordinarily beautiful that January. The calm that prevailed in the mornings, particularly since the loggers had left, was remarkable. The large bird-nest ferns with leaves as tall as a man rustled with the slightest breeze. The small *Flacourtia* saplings demurely yielded as I bent them out of the way, and near the coast, *Callophyllum* trees provided a buffer between the forest and the lava. Some days I would sit on the moss and stare upward at the majestic banyan trees soaring high above, each bearing a ton of epiphytic plants affixed to the branches. I listened to the cries of the birds and the occasional chatter of flying foxes. As I lay on the ground's dense carpet of moss, I sensed the forest's embrace.

The forest, of course, is not sentient, but an assemblage of rock, dirt, ferns, trees, lizards, birds, and flying foxes. But we are sentient, and it seems paradoxical that the fate of such an ancient forest is dependent on our whims. Acts of conservation by humans are ultimately ephemeral, and hence completely dependent on the unbroken continuity of our collective will to conserve. Thus we are faced with the unpleasant truth that there can be no irreversible commitment to conservation.

"Is there a permanent solution for the rain forest?" I'm sometimes asked. "Yes," I reply, "cut it down. Then it will never grow back and you won't have to worry about it again. But if you want to save it, each day you have to decide not to destroy it, and must trust that others will reaffirm that decision after you have left the scene. Since we will always have machetes, chain saws, and bulldozers at our disposal, we must collectively decide each day not to use them." Thus the indigenous people are correct in their view that tropical rain forest conservation is ultimately a spiritual issue. We are faced with an interesting existential paradox, the sort that Camus or Sartre might have written about: even though the forest doesn't care what we do and is supremely unaware of our existence, we remain both the sole means of its destruction and the only means of its salvation.

But agencies beyond human control play a crucial role as well. "This group [of islands], like all in nearly the same parallel of latitude, is subject to violent hurricanes between the months of November and May," John Erskine wrote in 1849.

> These hurricanes, called by the natives "afa fuli fao," or knock-down winds, seem to be rotary storms of small diameter, passing in general from a westerly to an easterly point. . . . The last had occurred on the 26th of December, 1844. . . . The sea was said on this occasion to have been heavier than any of the natives remembered to have seen it, and to have raised several islands of dead coral on different parts of the reef, which were pointed out to us as still existing.[57]

On the 5th of April 1850, sixteen months after the above was written, Apia was nearly destroyed by a hurricane, the iron chapel being the only building uninjured. Three vessels were wrecked in the harbor.

There was similar destruction of shipping by a hurricane in March 1889. An American and a German naval squadron in Apia harbor were both completely destroyed. Since the warships were on the verge of hostilities over the political status of the islands, their destruction was considered providential by the Samoans. Yet what gun and storm failed to accomplish, the stroke of a pen established in the Berlin Treaty of 1899: Samoa lost its sovereignty, and was partitioned between the United States, which took the eastern part of the archipelago, and Germany, which seized the west. At the beginning of World War I, Western Samoa was seized from Germany by New Zealand, which administered it until it gained independence in 1962.

Though hurricanes have had significant political, agricultural, and ecological consequences for Samoa, they have not typically resulted in catastrophic loss of life. During a storm in 1900 in Galveston, Texas, over 6,000 people died, and during the 1928 hurricane in Florida, 1,836 perished, largely due to high seas. Throughout the world, 90 percent of all hurricane-associated deaths are caused by storm surges.[58] Strong winds during a hurricane can create a gigantic dome of water that, if coincident with high tide, can raise sea levels over twenty feet. With the exception of a few places such as Falealupo, Samoan villages are usually protected by fringing reefs and high ground from such wave action.

The last week of January, I flew from Asau to Apia to greet some of the visitors arriving for the anniversary of the covenant. To my surprise, they brought with them a large cardboard box filled with antibiotics that the Phillips family had sent down to us. The plan was for me to fly back to Savaii the next day, with the other donors arriving for the ceremony two days later. But the plane didn't take off the next day because of stormy weather, and on the following day the airstrip was still closed, so I caught a bus to the Mulifanua ferry and rode the boat across. The sky was very dark and the sea was rough. Many of the Samoans on the ferry recognized me, so I was able to huddle on a bench and listen to their stories about fishing and seafaring.

Soon I felt a tap on my shoulder. "Excuse me, but do you live here?" a young *palagi* man with a large backpack asked. I grimaced inwardly—adventure tourism had arrived in Savaii.

"Yes, I do live here. What can I do to help you?"

"I was wondering if you might know of any place we can stay," he said, gesturing toward his three companions. Any Samoan would have readily volunteered their *fale* to the group, but even though I myself was the recipient of Pela's hospitality, I couldn't bear the thought of having four new *palagi* descending on me at this crucial time.

"There is a very good but inexpensive hotel at Safua run by Moelagi Jackson," I explained. "Just take a Salelologa bus at the ferry terminus and they'll drop you off." The backpackers obviously wanted to talk with me, but a continued conversation in English would have excluded my Samoan friends, so I excused myself and returned to chatting with the Samoans.

Recently more *palagi* had started showing up at our *fale* in the village, as if we were an outpost of Western culture, like a sort of white trader's family in the old West. Such visits were often awkward, since we preferred the company of the villagers: conversations in English precluded most of our village friends from participation. Besides, while in Samoa, we wanted to be with the Samoans. But my avoidance of the white expatriate community in Apia extracted a toll on my reputation. As word of my activities in Falealupo spread, we became the subject of cruel rumors. Some claimed that I had brought a beer truck into the village and got all the chiefs drunk to sign the covenant. Others asserted that despite the establishment of the preserve, the sawmill had set up operations in Falealupo and was busy cutting down the forest. And others professed that my conservation efforts were merely a ruse to convert the village to Mormonism.

From the ferry deck, I saw that we were approaching the familiar wharf at Salelologa. As I looked at the backpackers, I wondered if the Falealupo Rain Forest Preserve would attract more adventure tourists to Falealupo, and what the ultimate impact of tourism on the village would be. When our family first came to Falealupo, we were the first *palagi* people to have ever spent more than a few hours there. And now, I realized, the rain forest preserve would be showing up on tourist maps and in guidebooks. Had I really helped protect Samoan culture, or had I merely hastened its dissolution?

After traveling by bus to Falealupo from the wharf, I was surprised to see ropes tied across the roof of our *fale*. Our vehicle was nowhere in sight.

"What happened?" I asked Barbara.

"The night you left we had some very high winds and a storm. Lamositele got worried and moved us and our stuff out of the *fale* and up inland to Pela's house. What was the weather like in Apia?"

"It was stormy, so the planes were grounded. I returned to Savaii by boat. What happened to our van? Where is it?"

"Lamositele was worried that the trees might fall on it, so he had me drive it inland."

"But it wasn't a bad storm, was it?"

"No, Lamositele just wanted to play it safe. The whole sky was gray in the afternoon, and he said that he didn't like the look of it."

I was touched by Lamositele's concern, but believed that once again he was being overprotective. I went into Pela's house and saw her husband lying on the mat with his eyes open. "Lilo—how are you?" I asked him.

There was no answer, so I took his frail, aged hand in mine. "Lilo—it's Koki," I said.

Lilo looked up at me. "Koki, Koki," he said softly with a slight smile on his lips. I stayed with him a moment and then went out to visit with Barbara.

"I'm really concerned about Lilo. He seems to be fading fast. He hardly recognized me," I said.

"I know. He's so old. He must be nearly ninety. He's a wonderful old man," Barbara said softly.

Barbara opened the carton from Lee and Kristy and was astonished to see plastic bags containing thousands of pills, principally penicillin and tetracycline.

"Did you ask them to send down any antibiotics?" I asked.

Barbara shrugged.

"What can we ever do with all this medicine?"

"Maybe it will come in handy," Barbara offered.

The Swedish students arrived several days later amid the bustle of preparations for the one-year anniversary of the covenant. Rex Maughan, Jim Winegar, and Dan Wakefield were in Apia and ready to come to Savaii. As I awoke in the morning, the villagers were preparing pigs for slaughter and decorating the village huts. Each house post was wrapped in woven coconut fronds interlaced with red torch ginger and white *Alpina* flowers. The sky was overcast, and it was a bit windy, but I was hopeful that it would be clear and sunny by the next day.

The Swedish students, eleven in all, arrived promptly at 9:00 A.M. at our *fale* in a rented white van. They brought a surprise: John Cannon, a BYU student who had taken my evolution course and had developed an interest in coming to Samoa.

I assured John that he would be welcome to stay in our *fale*, and then introduced John and the Swedish students to our Samoan family. Since the Swedish students came in their own rented van, we drove inland to the forest with half of the students in my vehicle and half in their van. I decided to show them the cloud forest high above the logging areas at Asau. We drove up to the main road and on up to Vaisala, where we stopped at the little store to purchase cookies and soft drinks for lunch. We then started up the winding logging road toward the center ridge of the island. At the top, we stopped in a clearing. Several of the Swedish students who were expert at ornithology started looking for birds. They called to me with excitement because they had found a species they hadn't seen in the lowlands. I was slow to respond, however, because I had spied a species of the endemic tree genus *Sarcopygme* that I hadn't collected before. It was a beautiful tree with large, wide, undivided leaves and a long, pendulous flowering head that looked like a snowball. I had difficulty collecting samples, though, because I had forgotten my machete and had only my plant snips.

In my excitement, I had failed to notice that the skies had darkened and the wind had picked up significantly. By the time I got back to the van, the branches of the trees were bending in the wind. We piled into the vans and started back down the dirt road. After traveling about half a mile, we were stopped by two trees that had fallen across the road. The students and I were able to push the first tree to the side, but the second one, a little less than a meter in diameter, was too heavy.

A few of the students looked concerned, but their frowns turned to laughs when I pulled out my trusty Swiss Army knife and started sawing. Soon they saw that I wasn't kidding—this crazy American really was going to try to cut the tree in half with the blade of a pocket knife. In ten minutes or so I had worked up a sweat, but had created a large gash in the tree. One of the Swedes grabbed the blade and took a turn, and in about half an hour we had a large cut completely encircling the tree.

"We'll never get through it with that," one of the students said.

"True enough," I answered, "but I'll bet we have significantly weakened it. Let's all push on it again."

With a heave we managed to break the tree trunk along the cut line. We then slid the two pieces of trunk off the road.

Rain had started by the time we arrived at the main road at 2:00 P.M. I encouraged the students to drive directly back to the Safua hotel on the other side of the island. They climbed into their van and left. When I returned with John to Falealupo, Lamositele and several other men were still trying to finish decorating the huts for the next day's celebration. John was tired and, I suspected, suffering from a bit of culture shock, so I encouraged him to take a nap in our *fale*. I told Lamositele that I was going to go to the rain forest for an hour or so.

"Why?" he queried me. "It looks like we're in for a big storm."

"I want to go to the end of the trail that we built and mark another half mile or so for trail construction, at least to the top of Mount Mulimauga. I think if we show the foreign donors the marked trail, they might donate funds to the village to pay for further improvements for the preserve."

Lamositele refused to let me go alone, and so together we jumped into the van and drove off. In a few minutes we arrived at the trailhead, which was marked with a large sign the village had erected. We walked quickly down the trail. The village had designed the trail to meander through the forest, and at one point it was even threaded through the stilt roots of a banyan tree. About a mile of the trail had already been completed, and in areas that were particularly rocky, the villagers had carried in sawdust from the logging scar to smooth it. We arrived at the end of the trail, and I pulled some fluorescent orange flagging tape from my rucksack. I took a compass bearing from the top of a log where I could see the summit of Mount Mulimauga.

"Let's go," I said to Lamositele.

We climbed through the forest, tying pieces of the flagging tape to the trees as we passed. The sky was overcast, and there was intermittent rain. A few small branches had blown to the ground, but the wind velocity was less than what I had experienced earlier that day in the cloud forest. As we walked, Lamositele and I tried to visualize where a trail might be built, plotting our path around small ravines and large trees. Since no previous trail had been cleared, the going was tough. With an eye to the weather, both of us were moving as quickly as we could, but it took us the better part of an hour to make the half mile or so to the base of the mountain. Despite the storm, it was muggy, and Lamositele and I were both sweating.

"Take a rest?" I asked.

"Why don't you stop," he said, "and I'll go on ahead."

I sat on a log for a few minutes to catch my breath. Occasional gusts of wind shook the rain off the tree limbs far above me. Otherwise, the forest was strangely silent. I couldn't hear a single fruit pigeon or honeycreeper or flying fox. I rechecked the compass that was hanging around my neck and called out to Lamositele.

"Up here!" came a voice from the mountain. I started up the steep slope, marking possible trail switchbacks with the flagging tape. Soon I had caught up with Lamositele, and together we climbed the last several hundred meters to the summit.

The top of Mount Mulimauga was, as might be expected of an old volcanic cinder cone, flat. We walked to the south crest, where we could see out over the entire Falealupo forest. In the far distance I could see the road where I had left our vehicle, and beyond, the beginning of the village plantations. Below us the rain forest canopy stretched like a green carpet, punctuated by the occasional banyan tree that towered majestically above it. Downward from the banyans dangled long woody lianas, some as thick as the steel cables that support suspension bridges. Far to the west, I could see the ocean beneath the dark storm clouds. The view was spectacular. The village had indeed chosen well when they decided to build a backpacker's hut on this spot.

Lamositele grew impatient. "Nafanua," he said, "I think we better go back now before the storm grows worse."

Nodding, I took one last look out over the Falealupo rain forest. We returned quickly, following our flagging tape rather than the compass. At one rocky ravine we stopped at a tree I had marked, but couldn't see the next bit of flagging. Lamositele ventured ahead and then called back. A large tree trunk had obscured the flagging from our view. We arrived, wet and tired, at the vehicle at about 4:00 P.M.

When we got back to our *fale,* I took off my wet, muddy clothes. Putting on a swimsuit, I walked inland to the concrete water tank next to Pela's house for a shower. Filling the brown plastic bucket from the tap, I dipped an old corned beef can into the top, and ladled the cold water over my head. After stopping to soap up, I poured more water over my head. I was cold, but knew that when I put dry clothes on I would quickly warm up. Looking through the louvered windows into the European-style house where Pela and the older Lilo lived, I could see Pela sitting on the mat, smoking a cigar she had rolled herself, listening to the battery-powered radio.

"Pela, is the weather report on?" I asked.

"No, Nafanua," she replied. "But they said earlier that there might be a storm."

I frowned. If it was stormy, the little bush plane from Apia wouldn't be able to land on the Asau airstrip. Rex Maughan and the others wouldn't be able to fly over the next day for the ceremonies.

"Any word on the ferries?" I asked.

"Nothing," Pela responded.

Perhaps Rex and his group would be able to come over on the *Lady Samoa* if the plane wasn't flying. All of his group knew their way around Samoa pretty well, so I decided not to alert the village that there might be a delay in the ceremonies. I put on clean underwear, a T-shirt, and a freshly pressed lavalava. The dry clothing made me feel warm and comfortable.

Dinner was served, with all of us sitting on the floor as usual. Tita, Lamositele's eighteen-year-old daughter, had cooked a delicious soup with corned beef. Baked taro, reef fish, and some Saimin noodles completed the dinner. She had worked hard to prepare a nice meal, perhaps since John Cannon was with us. As might be expected on someone's first night in Savaii, John seemed a little dazed as he ate his dinner while trying to sit cross-legged on the mat. But he was polite and appreciative. After dinner, we lit the kerosene lantern and I read aloud from Roald Dahl's *Matilda* to the children.

Suddenly Pela appeared. "Nafanua," she said, "I think you should have John sleep up at our house. It looks like there will be a storm. Have him carry his stuff up to our house. You and your family ought to come as well."

"No thanks, Pela, but we'll send John up," I replied. "Perhaps he'll be more comfortable there."

Pela was always solicitous of the welfare of our guests, so I helped John carry his large rucksack up to her house. After seeing that John was settled, I walked back seaward to our *fale*. I switched on a small battery-powered radio. The national radio station, 2AP, broadcast a military march, some country music, and an old selection by the Bee Gees before finally giving the weather report.

"We alert our listeners that a storm advisory is in effect for Samoa, Tokelau, and Rose Atoll," the announcer said in English. "Rain and high winds are expected tonight and tomorrow with high seas. Fishermen and travelers are advised to take reasonable precautions."

Storm warnings of this nature are not uncommon in Samoa, but just to play it safe, I decided to listen to the weather report in the Samoan language. Even though it was just the second day of February, well before the hurricane season began in March, I wanted to make certain that there wasn't a hurricane or a typhoon offshore. When the storm advisory was broadcast in Samoan, I listened very carefully. Would the word *afā*, meaning "hurricane," be used? *Afā* is an interesting Samoan word. If there are no problems, Samoans say *"E le afāina,"* which means "Don't worry," but translates literally as "There is no hurricane."

I was beginning to worry, but the word *afā* wasn't mentioned once. Listeners were advised that there was a possibility of high winds, but there was no indication that inter-island shipping would be halted, so I was hopeful that Rex and his group would be able to catch the morning ferry over to Savaii.

We knelt together on the mats for our family prayers and then hung the mosquito nets over our sleeping mats. Barbara and I were in one mosquito net, and the children in another. Since it was starting to rain, we had lowered the *pola* blinds around our *fale*. I turned the knob on the Coleman lantern, and the light was slowly extinguished.

I woke to the sound of Mary crying. It was about two in the morning. "What's wrong?" I asked Barbara.

"Mary says that she's wet," Barbara answered.

"Isn't Mary a little old for that sort of thing?" I said with some irritation.

"It's probably Hillary who's had an accident."

"Well, I think Hillary's a little old for it too."

I shined the flashlight over on their mosquito net to make sure that everything was all right. Barbara pushed one of our dry sheets over to them and said, "Don't worry, Mary. Here's a dry sheet. We'll take care of everything in the morning. Go back to sleep."

I swung the flashlight back to put it under my pillow, but just as I turned it off, it illuminated a soccer-ball-sized chunk of brain coral that lay dripping at the edge of the hut near our sleeping mat. Odd. But I shrugged it off and rolled over to go back to sleep.

But sleep wouldn't come. Something wasn't right. Finally I sat up in the dark and tried to figure out what was bothering me. Of course I had been irritated at Mary or Hillary wetting the bed. Maybe I was concerned that I had spoken to them harshly, and I vowed to apologize the next morning. But still I reached under my pillow and pulled out the flashlight. I turned it on and shined it around the darkened interior of the hut. Its light played about the room until I saw, less than a meter from our heads, that large piece of brain coral again, dripping with seawater.

"Whatever is that doing in here?" I groggily wondered. "The sea is over three hundred meters away. I didn't bring it here, and the children know better than to break off any pieces of coral."

I was still contemplating the mystery of the coral when the next tidal wave hit our *fale*.

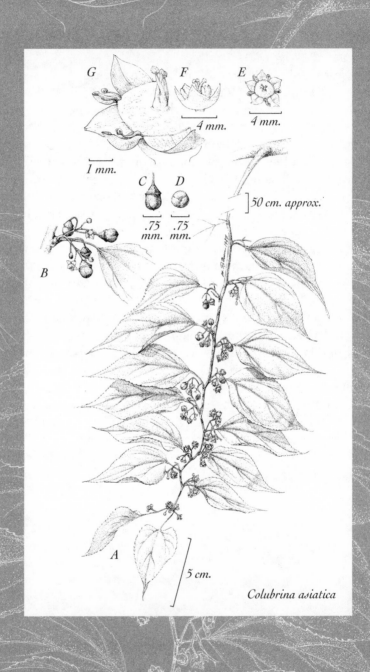

G

F

4 mm.

E

4 mm.

1 mm.

C

D

.75 mm.

.75 mm.

B

50 cm. approx.

A

5 cm.

Colubrina asiatica

CHAPTER 13

Deluge

Who hath desired the Sea?
The sight of salt water unbounded.

Rudyard Kipling,
"The Sea and the Hills"

Water gushed through the *pola* blinds into the *fale,* soaking our bedding. Barbara turned on our fluorescent lantern, and I glimpsed two more volleyball-sized chunks of coral that the wave had thrown into our hut. I separated the blinds and peered out toward the sea. It was dark and overcast, and the wind was blowing. I could see and hear the seawater receding from around our hut.

"A wave came all the way up here to our *fale!*" I remarked to Barbara, more curious than concerned. "Imagine that!"

Flashlights were also coming on in Lamositele and Fa'asaina's *fale* next door. Soon Fa'asaina came over.

None of our family had been frightened by the wave incursion. It was just a novelty.

Fa'asaina did not share our delight, however. Instead of smiling like the rest of us, her face was ashen and somber.

"I think you all better move up to Pela's house. I don't think it is safe here," Fa'asaina said.

Lamositele, who then entered the hut, shared her concern. "I really think we better move you up to Pela's house, Nafanua."

"But you said that a wave has never come this high before. Surely it won't happen again. Besides, I hate to disturb Pela and Lilo," I protested.

Just as I finished the sentence, Pela appeared in the hut in her nightdress. "Nafanua," she commanded, "bring your children. I want all of you to come up to my house *now.*"

One does not argue with Pela, so I meekly began collecting our wet bedding. It was dark and raining outside, so Barbara and I carried Hillary and Mary on our backs the hundred meters inland to Pela's house. Tita and Pela spread some sleeping mats on the concrete floor and laid our children on them with some dry bedding. I strapped on my headlamp and walked back to our *fale.* Lamositele caught up with me.

"Perhaps we should move all of your stuff up to Pela's," he suggested.

"No, let's wait until morning," I said. "I'm sure the storm will clear by then."

I looked around the hut at my research gear. I placed some of the electronic equipment high on the plant drier just in case another wave came. I stacked our suitcases on the little table and began to fill them. The antibiotics in the cardboard box that our friends had sent caught my eye. I pulled out the large waterproof box filled with silica gel that protected our photographic equipment against humidity. Working quickly, I replaced the cameras and night vision device with our passports, return airline tickets, traveler's checks, the Katydyn water filter, and our medical kit. I then packed the container with as many doses of antibiotics as I could. I sealed the box shut, making sure the rubber gasket was tight. "Just in case," I thought to myself. I threw the waterproof box into a net bag that I strapped onto my back like a rucksack. With one hand I grabbed my briefcase with my laptop computer. With the other I grabbed the large Pelican waterproof case with my camera gear, and headed toward Pela's house and another attempt at sleep.

I was up at 6:00 A.M. the next morning, 3 February 1990. The sky was dark and overcast. A gale-force wind was blowing from the sea. I walked out toward the beach. Waves broke constantly above the beach line, with a few big ones occasionally surging across the sandy road toward our *fale.* Lamositele was already up, packing up the belongings in his hut.

"I think we better move all of our stuff up to Pela's house now. Do you need some help with your things?" he asked.

"No, I think I can do it," I said.

Even though I had declined assistance, I was delighted when Tasi, Tive, and Folau, all teenage grandsons of Pela, showed up to help carry my research gear inland. With them lugging the heavy suitcases and boxes of books, we were able to transfer all of our things to Pela's house in a couple of trips.

Lamositele motioned toward our four-wheel-drive vehicle behind our *fale*. "Maybe you should move your car to higher ground up by the Catholic minister's residence," he said, pointing to a rocky knoll.

"No, I think it will be O.K. next to Pela's house," I said. I jumped in and drove to the front of Pela's *palagi*-style house between the breadfruit trees. It was well away from any of the large *Erythrina* or coconut trees that might conceivably fall on it if the wind got really bad.

By 7:30 there was sufficient light for me to take some pictures. I grabbed my video camera and walked down to the front of our *fale* to film the waves breaking over the road. The wind velocity was obviously on the increase. Salt spray and rain droplets filled the air, making it impossible to see the reef. I searched for a protected spot and even then had to wipe the spray several times from the lens of the video camera. I had gotten only about thirty seconds on tape when the recording mechanism stopped. The tape was jammed. I returned to Pela's house to retrieve my still camera. I stopped to put on my swimsuit, still wet from last night's shower. I grabbed one of my 35 mm cameras, loaded some fresh film into it, and hurried back to our *fale*. The children wanted to come, but I asked them to stay in Pela's house.

The waves were now surging up to the foundation of our house. I was acutely aware that I was witnessing something important— the highest wave incursions in Samoa's recorded history—and so was eager to have a good photographic chronicle of the events. Even though the wind velocity had steadily increased, I still didn't sense any real danger. Lamositele and one of his sons still were trying to fasten flowers to the meeting house posts for the covenant celebration. I went over and shouted above the gale, "Lamositele, there's no way the donors can make it to Savaii with these high seas. Let's forget it."

I started snapping pictures of the waves breaking with full force across the road and the tops of the coconut trees bending sideways up to a meter in the wind. Lamositele interrupted me briefly to help him carry Fa'asaina's sewing machine up to Pela's house. By this time the waves were regularly passing right through our *fale*. In effect, the sea had now risen to the height of the road, which stood three meters above sea level.

I constantly had to wipe salt spray from the lens, but I was getting some exciting shots. What appeared to be a three-meter wave began to surge in the distance, so I stood with my feet firmly planted in Lamositele's *fale,* watching the wave through the camera viewfinder. It was already over the road and rushing through our former residence when I suddenly realized that the wave would hit me with full force. Dropping the camera by its neck strap, I leapt as high as I could in the *fale,* seizing one of the rafters. The wave crashed through the hut like a freight train, nearly a meter and a half above the floor. Just as it hit, I pulled my legs up onto the rafter. The wave carried with it fallen trees, large stones, and coral heads, and the little *fale* swayed with its force. After the wave retreated, I dropped down from the rafter and waded back to Pela's house. The photography session was over.

As a family, we huddled in Pela's house, peering seaward through the louvered windows. The children were now becoming frightened, and I was growing very concerned. I went out the back door to scout out a possible escape route, just in case. Three hundred meters behind Pela's house and about eight meters higher in elevation was the house of Mariana, who had recently left to visit her children in New Zealand. Between Pela's house and Mariana's was a small depression containing a path marked with large lava rocks between the coconut and kapok trees. I walked halfway to Mariana's house, hoping to familiarize myself with the path. The rain became torrential, so I returned to Pela's house.

It was about 8:30 when the waves began to hit my car. As soon as they receded, I ran out and jumped in to try to drive it to higher ground. But while I was trying to start it, another wave hit the car, twisting it around and driving it into a breadfruit tree with a sickening thud. Water started pouring in through the crack in the door, and for a moment I thought I would be trapped. But the water receded again, and I was able to escape and wade back to Pela's house.

The sea was now already up to the doorstep of Pela's house. She quickly closed and bolted the door after I entered. All of us were together: Pela and Lilo; Lamositele and Fa'asaina; eight grandchildren of Pela and Lilo; Barbara and I and our four children; and the student from BYU, John Cannon. We hardly spoke as we gazed at the ongoing destruction outside the front window. I could see the little shed by the side of our hut, twisting and twirling in the waves. Then we saw our *fale,* with waves crashing through it, sagging on its side. The wind's velocity increased even more, and it began to sound like a long, low whistle, which forced us to shout to be heard. But inside Pela's house we were silent, transfixed by the scene outside.

I glanced at my watch. It was now about 9:00 A.M. Suddenly Tita screamed. Through the louvered window I could see a three-meter-high wave headed straight for us. We jumped to grab our children. When the wave hit, it shattered the glass panes, filling the house with mud and broken glass. Somehow, miraculously, Pela's front wall, groaning under the pressure of water and debris, did not cave in.

I spoke rapidly in Samoan to Lamositele. "Quick! We've got to get out of here before the next big one hits! Let's go out the back, up to Mariana's house. I've scouted the route and it's clear. You and your boys take care of Lilo and Pela, and we'll get our children."

Lamositele nodded and began speaking quickly to Fa'asaina and his boys. I grabbed the net bag that held the waterproof case with our passports and antibiotics and strapped it onto my back. I turned to John Cannon. "John! Carry Mary. I'll carry Hillary."

I turned to Emily and Paul. "Leave everything and hold hands. Don't let go of each other for any reason. Stay right next to me. We've got to leave. Now. Out the back. Follow me."

I didn't have to say anything to Barbara because I knew that as always she would do what needed to be done. Out of the corner of my eye, I could see Lamositele's son Lui and his nephew Keli lifting the eighty-six-year-old Lilo. Lamositele helped Pela, and Fa'asaina clutched the terrified fourteen-month-old Sapa to her breast.

We went to the back door, which was usually about a meter above ground level. As we opened the door, water poured over the doorsill. I looked at Barbara. "We've got to keep the children together."

John Cannon struggled along with Mary on his back piggyback-style. As we left Pela's house, I noticed the inflatable vinyl raft that Mary had received for Christmas. I looked at John. "Can you pull Mary on the raft?"

He gave me the thumbs-up sign and took the raft. Just then the next wave hit the front of the house with a thud. "Let's go!" I yelled, stepping out the back door into the water.

My feet couldn't find the rocks that I had previously scouted. Everywhere there was sea foam and floating debris. With Hillary holding on tight, I swam for about five meters and then got my footing. As I was swimming, I was gripped by an irrational fear of encountering a floating centipede. Ever since one bit me inside the ear, I have had a phobia of centipedes. There were already dead pigs and other domestic animals floating in the water. Once I had secure footing, I turned around and saw that Barbara, Emily, and Paul were right behind me, with John carrying Mary on his back while pulling her little raft. We all waded another thirty meters through the coconut trees, pausing to climb over a fallen but submerged trunk, and then climbed the hill to Mariana's house.

Samoa, the son of Mariana, and his wife Lologo, Seumanutafa Siaosi's daughter, greeted us and handed us towels to dry off with. I turned around to see the boys carrying Lilo in. Lamositele and Pela followed. Pela was in tears. "You've got to try to save some of our things," she sobbed to Lamositele. "You've got to save our photographs."

Samoa and Lamositele took Pela and Lilo into a private room. We watched the storm for half an hour. Not seeing the waves rise any farther, we decided to try to retrieve some of our possessions from Pela's house. Swimming back and forth, and using Mary's little raft as a floating barge, Lamositele and I, with the help of Pela's grandsons, Tasi, Lui, Keli, and Folau, were able to carry all of Pela's personal belongings, her photos, cookware, and dishes, as well as all of our research gear, to Mariana's house.

My car lay twisted with the tires up near the side of Pela's house. All of the village seaward from her house had now been washed out to sea. I remained confident, though, that we were safe at Mariana's, since it was the highest dwelling in the village. But the sea continued to rise—the water depth was now over a meter in Pela's house.

Samoa and Lamositele scurried around his little house, which was now jammed with people. Using hammers, nails, and ropes, they tried to secure the roof, which threatened to blow away.

I asked Lamositele if there were any other houses farther inland. Perhaps I had missed some in my previous wanderings about the village.

His answer rang dead: "None. This is it: there is nowhere else to go."

I looked deep into his eyes. "We've got to have some alternatives, just in case."

He looked at me, pained, "There are no alternatives. This is the end of the village."

I went out the back of Mariana's house into the hurricane. Lamositele was right. There was nothing but secondary forest behind her house. I scouted a path toward it. But as I got closer, I heard the tremendous crash of trees falling and saw pieces of corrugated roofing iron flying through the air, like giant twisted razors. Nothing but death seemed to await us if we ventured into the forest—we would all be killed by flying debris or falling trees.

I returned to Mariana's house, shaken. It was now 2:00 P.M. Our children were huddled together with the Samoan children. All were frightened. Fa'asaina had gotten a kerosene burner to work in the back of the house and was heating water for cocoa and rice. We tried to get the children to play "Little Peter Rabbit" in both English and Samoan and then to sing some Samoan songs. The children participated, but with little enthusiasm.

The wind speed kept increasing. The water kept rising. La'ulu, a neighbor and brother-in-law of Lamositele's, suddenly burst in through the back door.

"How did you get here?" I asked.

La'ulu told us the terrible news. "The Avatā portion of the village is washed away," he said. "All of the houses, everything—all is gone."

"What about the villagers?" I asked.

"I don't know about the people in Salesau, Faleolo, and Vaotupua," he said. "But in Avatā and Malaetele, most of the people made it up to the school."

"Is the school building holding together?" I asked.

"The main building has lost its roof, but the lower floor is intact, as well as the south classroom wing."

"How many villagers are in there?" I asked.

"I don't know—maybe seventy or eighty."

I became very concerned. The total population of Falealupo was over a thousand, with about half of the village living by the sea and the others far inland near their plantations. Of the five hundred or so villagers living by the sea, less than a fifth had made it to the school. Over four hundred people might be trapped or dead.

"I better get back before it gets any worse," he yelled over the howl of the wind. "If the water rises any more, it won't be possible." I urged La'ulu to stay, as I thought we could weather the storm in Mariana's house. But he was insistent and ran out the back door. I didn't know if I would ever see him alive again.

By 4:00 P.M., the sea level was less than a meter below Mariana's house and well above the roof of Pela's house. At 5:00 P.M., Barbara and the children and I prayed, asking God to spare our lives and to protect Mariana's house from being washed away, for we didn't have anywhere else to go. I noticed that John Cannon had grown very quiet. He spent most of his time checking and rechecking the camera gear in his rucksack. I went outside again. The wind velocity had increased still more. With a giant crash, a large kapok tree fell to the ground, barely missing Mariana's house. It was growing misty and dark, but I could see some smaller trees flying through the air, along with pieces of corrugated steel and entire thatched roofs from destroyed huts. It was difficult in the fading light to discern the difference between the raging sea with its burden of rock and debris and the mist-filled air filled with flying timbers, tin, and tree limbs.

I went to the back room and found Lamositele, Fa'asaina, and their children chanting prayers from the catechism, asking each Catholic saint in turn to save them.

"Holy will of Jesus, send us your peace."

"Holy Mary, who is without sin from the beginning, save us."

"Saint Joseph, save us."

"Saint Falaniko, save us."

"Saint Peter, save us."

"Saint Teresa, save us."

I went again to the front door, and saw that the sea had now risen to the doorsill. I could see only a few meters into the mist, but far enough to see the next wave coming in time to shut the front door. It was no use. The dull brown water rushed through the louvered windows, quickly filling the little house. I ran into the bedroom and saw our children terrified, standing knee-deep in water. A second dull thud hit the house, and more seawater surged through it.

I turned to Barbara. "Our only chance to stay alive is to try to make it to the forest. It will be bad, but if we stay here, we'll all die."

Barbara nodded grimly. I felt to make sure that my net bag with the antibiotics was still strapped to my back. We stepped into the back room. For the first time in my memory, Lamositele, always so wise, seemed dazed and incapable of taking action. Lamositele was witnessing the destruction of his universe: everything he had known from birth, everything his parents had known and their parents had known. Never before in Samoan history or legend had an entire village been swept out to sea. Nothing that had happened in the past could guide us; all traditional wisdom was useless.

"Lamositele," I shouted over the wind. "We've got to leave. We've got to get out of here before the next wave hits. I've scouted the way. We'll go into the forest and then try to reach the school. We need to stay together."

Lamositele looked at me and softly said, *"Ua pō."*—"It's dark." I glanced at my watch. It was 6:30 P.M. We were all knee-deep in water. The children were too frightened to cry.

"Lamositele," I said. "We must leave *now,* before the next wave hits, or we'll die here. Let's go."

Lamositele nodded, and had Lui lift Lilo onto his shoulders. Tita took Silia and Susana by the hand. Lamositele was the last one out of the house, leading Pela. I saw Tasi and Folau grab some of our suitcases. "Forget them," I said. "Leave them here. We need to go _now!_"

Keli and Sulutolu led the way. We had to wade and swim a short distance, clamber up some rocks, and then stumble into the forest. With the wind velocity over 150 miles per hour, it was nearly impossible to stand up. Most of the tree trunks had snapped, but a few stragglers still stood, stripped of leaves and branches. The air was dark and filled with spray, but I could see tree trunks, metal roofing, lumber, and pieces of concrete water tanks swirling past. In one tree we passed, I saw a small piece of roofing tin embedded six inches deep in the trunk. Communication was impossible, unless we screamed into each other's ears. We stumbled across the rough black lava through some scattered trees.

"Heavenly Father," I pleaded silently, "please protect us so that we won't be hit by a piece of metal or a tree."

Lamositele's son Lui, with Lilo still wrapped around his shoulders, was just ahead. We followed behind, with Barbara, Emily, and Paul in the lead, and John Cannon carrying Mary. I brought up the rear, carrying Hillary. Lui set Lilo down on a rock, and we all stopped for several minutes while Folau and Tasi led the way into what had once been a thicket. Pausing in the forest was as difficult as waiting on a battlefield. Every second of standing in the open exposed us to death from flying debris. After trying unsuccessfully to negotiate the thicket, Folau and Tasi came out, motioning that passage was impossible. I approached Lamositele and shouted above the wind into his ear, "Let's try to get to the school, but if we can't, we've got to go farther into the forest away from the waves. Perhaps we can stay alive up there." I dimly envisioned us digging into the bare ground, or trying to cover the children with rocks and our bodies to protect them from the flying debris. But it was almost dark, and I didn't know if we could crawl much farther inland with the children.

Lamositele put his mouth to my ear and yelled, "Nafanua, I know another way to the school. It's a longer way, but maybe we can make it. Follow me!"

Lamositele started to lead us around the thicket and through a tangled piece of forest. We were slowed when we had to negotiate some tangled barbed wire from an old fence. I felt Hillary's grasp around my neck loosening. I was terrified that she might let go and be lost. Already I could see concrete water tanks tumbling end over end across the lava. I screamed at her, "Hillary—you've got to hold on! Don't let go."

I despaired for a moment, and then felt her tiny hands tighten around my neck. That little gesture gave me the first hope that we might not all die. I went up to John and yelled above the screaming wind to Mary: "Hold on! Hold on tight! We're going to make it."

We followed Lamositele for what seemed an eternity. Then I began to sense that we were heading back toward the sea—and back toward the school. We stepped through a glen of shattered trees, and finally saw the silhouette of the Falealupo Primary School, the building that we had sacrificed so much for. Half of the roof was already gone, and the other half was spinning in the wind, ready to break loose at any minute. The school was surrounded by the sea. Swirling water lay between us and safety.

Several villagers were huddled at the school, including Chief Soifua Manuele. He saw us trying to make it and came from the other side to help. Fortunately, the water was only waist-deep, and soon we had all made it across.

The school building was jammed with wet and terrified villagers from Avatā, Malaetele, Salesau, and Faleolo. Only the villagers from Vaotupua and a few from Avatā were unaccounted for. We were led to one of the few rooms that had a roof remaining and put down our children. Susi La'ulu, Pela's married daughter, ran and embraced Barbara in welcome. Somehow, miraculously, not a single member of our party was injured. Lilo was unconscious, but breathing. We immediately laid him on a mat. I took the net bag with the antibiotics from my back and set it down by Barbara, who was comforting our children. I was shocked to see Tasi and Keli carry in two of our suitcases, even though I had told them to leave everything in the house.

I looked around, grateful for a moment of respite, and then realized that I had left my computer and my notebooks at Mariana's house. The machine meant little to me, but the years of ethnobotanical research recorded in the computer files and in my notebooks—everything I had learned from the Samoan healers about potential cures—would be destroyed. I looked at Barbara. I knew that she would refuse to let me return. And I knew deep inside that it would be a stupid and foolhardy act to risk my life for those notes. I looked at Lamositele. He was tending to his mother, who was in shock. He would physically restrain me from going, or worse, try to come with me, risking his life as well. I had to act quickly, before anyone noticed. I quietly walked out of the room, up to the sea behind the school, and leapt in.

I waded across the water and entered the glen. It was very dark, and I was all alone. It was difficult to keep my bearings, but I thought I could make it. Without having to carry a child, I found I could move more quickly, although during high gusts I had to drop to the ground to stop from being blown away. I went through the glen, through the barbed wire, across the secondary forest, and then started seaward along the lava. Somehow, I miscalculated and missed Mariana's house. I had just clambered over a downed tree when I saw the crest of a wave coming toward me. I grasped the downed tree trunk, held my breath, let the wave sweep over me, and then started swimming for my life.

My legs got tangled in some debris. I felt the thorns of a sub-merged vine gouge my legs and a floating mat of *fisoa* bush entwine my torso. I held my breath again and tried to swim through the next wave to a coconut tree. At that moment I knew I would die. Every-thing seemed to move slowly, and I was deeply overcome by a pro-found sense of how foolish I had been and how I was letting my children down.

I reached out and suddenly felt the smooth trunk of the coconut tree. I pulled myself out of the water and alternately ran and swam in the direction of Mariana's house. It was awash with chest-deep water. I waded in and found the briefcase containing my notebooks and laptop computer high on a dresser. The water had almost reached it. Floating nearby, at chest height, was the Pelican case with all of my camera gear. I grabbed both cases and ran straight back into the forest. I ran as fast as I could, praying that I could stay alive. I went through the grove, through the barbed wire, waded through the water, and burst back into the school.

Barbara ran toward me. "Where have you been! Someone saw you go back into the water! We thought you were drowned!" Then she saw the cases in my dripping hands. "You went back for your computer and notebook? Paul—that's stupid! You could have been killed!"

I looked at Barbara sheepishly. "I know, and I shouldn't have done it. But at least I'm back now."

Lamositele came up. "I started out to look for you," he said. "Lilo called for you, but we couldn't find you. We were afraid you had drowned."

"I'm sorry," I said.

I walked with Barbara to the seaward side of the school build-ing. The waves were already breaking on the foundation. We turned to go back into the schoolroom, when suddenly a monster wave hit, crashing into the rooms and flooding them with water. We were already ankle-deep in water when another wave hit, increasing the water depth. We stacked the school benches on one side of the room and lifted the old people and the children up as high as we could. I waded out through the water and looked at the concrete foundation of the school. It was eroding away. Soifuā and Lamositele ap-proached me.

"Nafanua, all the villagers want you to lead them in prayer to plead with God to spare our lives. Please come."

The schoolroom was packed with people. We knelt in the water. I began praying in Samoan the same way I had been taught by my mother to pray in English so many years ago.

"Heavenly Father: We are in awe of Thy works and acknowledge to Thee Thy power and majesty. We ask Thy forgiveness of our sins, and pride, and vanity, for we know that Thou rulest the world in glory. Heavenly Father, we kneel not as *palagi* and Samoan people, but united as Thy children. We have lost nearly all of our possessions, and will gladly suffer the loss of our few remaining possessions if Thou desire it."

"We humbly ask for only one thing: if our lives are to be forfeited, wilt Thou please spare the lives of the elderly and the infirm, and please use Thy power to save our children, that they may live out their lives. We humbly beseech Thee to save us in the name of Jesus Christ, Amen."

The "Amen" of every villager echoed over the scream of the wind outside.

I took Barbara by the hand and walked outside. The school foundation began to crumble. It appeared that the school would be washed away like every other structure in the village. The school was completely surrounded by the sea. There was no escape.

I clasped Barbara's hand tenderly. "Barbara, I'm afraid this is it. We can't last. We will all die when the next wave hits."

Barbara pressed close to me. "Can't we help the children climb a tree?" she asked.

"They'd get blown off, or the tree would snap. Whatever is going to happen to our family will happen here. There is nowhere else to go. Thank you for sharing my life. Thank you for being my wife. Thank you for your love. I love you."

"I love you too, Paul." Her voice was barely audible as she squeezed my hand.

Barbara returned to the classroom to be with the children. I sat just outside out on a broken chair in the water, facing the sea, waiting for the wave that would sweep us away.

10 cm.

1 cm.

Erythrina variegata

CHAPTER 14

LOSS

*The true paradises are
paradises we have lost.*

Marcel Proust,
Time Regained

I sat in the chair, facing the sea, and waited. Futile though it was, I felt a deep responsibility to be prepared to raise the alarm when the next wave came, to alert the others to begin the last bit of frenzied activity before we were all washed out to sea. Unlike the other three structures where we had sought refuge that day, there was no escape route from here, no back path, nowhere else to run to. The Falealupo Primary School was the only building in the entire village that had not been swept away by the sea. But now it was surrounded by waves.

I sat with my ankles and legs in the water, facing the sea. Water had spread throughout the school. The only refuge from the sea was atop the benches we had piled up on one side of the room. I listened to the scream of the wind and the muted sounds of humanity behind me. Nearly three hundred terrified villagers had jammed into the school, fleeing the destruction of their homes, churches, and all that they had known. All of their material possessions had been lost, and, with the next big wave, even the small flame of life that each of us clung to would be snuffed out.

I waited and waited, and finally went back into the schoolroom. It was 2:00 A.M. Some of the other chiefs had organized a series of watches and urged me to try to get some sleep. A kerosene lamp someone had found in the school illuminated the room with a faint orange glow. Several benches had been put together and mats laid on top, forming a bed for my family and John. At the other side of the room, on a different island of benches, Pela's family slept. Much of the water had been mopped out of the room, but the floor was still very wet. I was exhausted. I wanted to sleep on the floor so that when the wave came, I would be the first to wake. I quietly found a mat and unrolled it. I wanted to touch Barbara one more time, and gave her hand a light squeeze. She awoke instantly, and seeing the mat on the concrete, left the children and joined me.

I lay down, and then heard a voice softly calling me: "Koki, Koki."

I sat up and listened. The howl of the hurricane outside was like an incessant roar. Then I heard the voice again, ever so softly, "Koki, Koki." It was coming from Pela's side of the room. I walked over, and saw that her husband Lilo was wide awake.

"Koki? Is that you? Are you O.K.?" Lilo asked me.

"Yes, Lilo. I'm here. I'm fine."

"Good," said Lilo. "I was worried about you earlier. You weren't here. I haven't felt this good for a long time. Something about the hurricane must have perked me up." Lilo then fell into a deep sleep.

I walked back to our mat and only then realized that this was the first conversation I had enjoyed with Pela's husband without the fog of Alzheimer's disease for over a year. I slept fitfully, waking instantly when airborne debris hit the school. With each impact, I listened to hear if a wave, rather than the wind, had propelled it. My clothes were soaked, so several times when I was jolted awake, I feared the wave had come.

I awoke at 6:00 A.M., surprised to be alive, and startled to see that many of the Samoans were up and about. The air was dark and thick. The loud, omnipresent whistle of the wind continued unabated, drowning out all conversation unless we screamed. I arose and went outside to check the foundation of the classroom wing where we were staying. The waves were still eating away at it, and large pieces of concrete had fallen into the sea, but no major wave had yet breached the building. As I looked at the main wing of the school, I noticed that during that night the roof had been blown totally away. Back inside the classroom, Susi La'ulu lit a kerosene burner for cooking. "How did you manage to get that here?" I asked.

"The night the storm began, La'ulu and I were afraid that our things would be damaged. Since we live next to the school, we brought several loads over. La'ulu even went down to Solia's guest house and brought up a box of canned goods. But then the storm got worse, so we all stayed here. And then our house was swept away by the waves."

I inspected the food Susi had been able to rescue. It consisted of several cans of beans, a few pounds of rice, ten sealed tins of peanuts and chips, and a few cans of fruit. Hardly enough to feed three hundred people. Soon Soifuā, the ranking chief present, came to me and said: "Nafanua, we're convening a chiefs' council in the principal's office."

About seven chiefs were already assembled in the small office. All of the chiefs looked toward me. "What should we do?" they asked.

"Do we have any deaths?" I asked.

"None here, but we've got only about a fourth of the village accounted for."

"Are there any serious injuries?" I asked. Nearly everybody in the school was scratched and cut up.

"Only one—a little boy whose arm was nearly cut off by a flying piece of tin," one of the chiefs offered.

Although Pela was one of Samoa's most gifted healers, the forest and beach medicinal plants she depended on for her practice had been destroyed by the hurricane. Unfortunately, with my store of antibiotics, I was the closest thing the village had to a doctor.

"I'll tend to him after our meeting."

We discussed together the need to ration the food and the remaining water in the school rain tank. The concrete tank had been punctured by a tree during the night and had lost most of its water. With the roof gone, we would not be able to catch more rainwater. We agreed to conduct an inventory of all food to ensure that everyone would be fed communally. In accordance with Samoan custom, at the onset of the disaster, all food resources became village property under the sole control of the chiefs.

I suggested that we send a small team into the storm to try to conduct an inventory of the village, so that we could estimate how many people were dead. Most of the villagers still unaccounted for lived several miles inland at their plantations and could not be reached, but they were probably safe from wave incursions. But another 250 people had, before the storm, resided by the sea in Vaotupua. Although they had probably perished in the waves, we couldn't discount the possibility that there might be small pockets of people left who were trapped or injured. I asked two of the chiefs to check Salesau and Vaotupua to the south of the school, and volunteered myself and Susi's husband, La'ulu, to check Avatā to the north. La'ulu looked at me with a pained expression. It was clearly difficult taking orders from a *palagi* chief.

"O.K.," I said to La'ulu, "We'll leave after I finish looking at the injured little boy."

I went back to the other room and found Barbara and the children awake. Emily and Paul asked if there was anything they could do to help. They could be cheerful and help keep Mary and Hillary happy, I suggested. In one of our soaked suitcases, they found a portable Ludo game, and soon had all the children involved in playing it. John Cannon seemed sluggish and lethargic. He was feverish, and I saw that his leg had a large gash in it, with red streaks going up toward his thigh. I opened, for the first time, the waterproof case I had strapped to my back the day before. Luckily, all of the antibiotics and drugs were dry and sound. I washed out John's infected gash, put Neosporin on it, and gave him 500 mg of penicillin. "Take two of these every four hours," I said to John, trying to sound like I knew what I was doing.

The mother brought the injured boy into a corner of the room that I had made into a makeshift infirmary. He was perhaps three years old, and was clutching his arm closely to his side. He looked up at me in terror. "Don't be afraid," I said to him. "I'm your friend."

He didn't want me to look at his arm. There was dried blood all over his shirt. Slowly, with much coaxing and prodding, we finally got him to unbend his arm. I was shocked at what I saw. It appeared that his hand had been nearly severed. I got some iodine surgical scrub out of the medical kit, washed my hands carefully with it, and then proceeded to wash the wound out. The boy screamed with pain and fear, but his mother held his arm steady for me to work on. As I cleansed the area, I could see that his hand was still attached to his arm, but the gash was deep, going to the bone. I rummaged in the medical kit, and was pleased to see a sterile suture pack. But there was no anesthetic.

I led the boy's mother aside and spoke quietly to her. "Your son's gash is really bad. He needs stitches if we are going to have even a chance of saving his hand. I'm not a real doctor, but if we can sew the wound up, perhaps when doctors get to us, they can help. But I don't have any anesthetics here—nothing to deaden the pain. Your boy will scream and cry when I start stitching, and you will have to hold his arm absolutely still until I finish. Do you think you can do that?"

The woman's eyes filled with tears. "I don't think I can stand to hear him scream any more. Can't you do something else?"

I looked at the boy's hand. He would probably lose it anyway, but it was worth an attempt to save it. I looked around the schoolroom, and found a flat piece of board from a chair seat. I asked Keli, who had found a machete, if he would go outside and cut the board into two small pieces. I then took some large Band-Aids from the medical kit, and with the scissors on my Swiss Army knife, cut pieces out from each side, so that the bandages resembled small butterflies.

"O.K.," I said to the mother, "hold him tight."

As the little boy screamed, I swabbed Neosporin all around the inside of the terrible wound. I then pressed the edges of the flesh together and applied the butterfly bandages. I used the two pieces of board as a splint to immobilize the hand, and finished by tying a sling around the boy's neck. I then cut one of the 250 mg penicillin tablets into four pieces with my knife, and had the mother help the boy swallow one.

"Have him take one of these pieces four times a day," I said. "And don't let him take that splint off. The only chance he has of healing the wound is to keep his hand immobilized. Just check his fingertips every hour or so, to make sure they aren't turning blue."

"Do you actually think he has a chance?" Barbara asked me later.

"No," I replied, "but it's the best I can do. Thank heaven there are no malpractice lawyers in Samoa. Even if we survive, I'd be thrown in prison for practicing medicine without a license."

Everything we owned, and everything in the school except for the antibiotics and medical kit, was wet. Dryness seemed to be an elusive quality, a memory of some previous condition that could never be restored. After sorting through my soggy and gritty research gear, I realized that it was time for La'ulu and me to make our village inventory. I removed my shirt and put on my swimsuit and tennis shoes. I split the drugs, packing half of them and the medical kit into the case. I checked the waterproof gasket, and then strapped the case to my back.

"Be careful," Barbara said.

In a feeble voice from the corner of the room, Lilo said in Samoan, "Koki—be careful!"

"Thanks, Lilo," I replied.

La'ulu and I stepped into the storm.

Between gusts of the hurricane, we ran from rock to rock, diving behind cover like soldiers under fire. During the gusts it was impossible to stand up, and I feared that we might become airborne. Even between gusts, we had to run leaning directly into the wind. The sand, propelled by the hurricane, blasted our skins, and our legs and chests oozed blood.

We ran behind Avatā, the part of the village that had been swept away, searching for any signs of life. The scene looked like the aftermath of a nuclear explosion. The Catholic church had disappeared beneath the waves; only a small portion of the crumbling front wall remained. We crouched together behind a tree trunk, waiting for a lull in the hurricane.

"Let's go back," La'ulu said. "There's nobody left alive."

"No, we've got to go on, just to make sure," I replied.

We started to run toward the remains of Pela's house when all of a sudden I heard La'ulu scream in terror at the top of his lungs, "Galu!" ("Wave!").

I looked seaward and saw a giant wave, bigger than any I had seen before, coming straight for us. La'ulu and I ran as fast as we could inland, but we were caught before we could make it to high ground. As the crest hit me, I instinctively leapt forward and swam with all my might, struggling to stay above the debris below. I plunged forward with the wave, almost like a body surfer, cruising inland over the remains of our village. I saw an erect coconut tree directly in front of me. I hit the tree with a bone-jarring thud, and grasped it with all my might. The wave swept over me, submerging me beneath the water, but miraculously I was not hit by tumbling rocks or logs. As the wave receded, I desperately clung to the tree, trying not to be dragged out to sea.

I looked around, fearing that La'ulu had drowned. Suddenly, I saw movement. A very bedraggled and terrified Samoan chief slowly pulled himself up from behind a tree trunk.

"We've got to get out of here!" La'ulu screamed at me.

I was inclined to agree.

We ran past the Catholic lay minister's hut. Miraculously, part of it was still standing, and we took shelter on its lee side. Over the scream of the wind, I thought I heard a voice from inside. I pounded on the side of the boards and then stopped. I heard a voice again.

"La'ulu!" I screamed above the wind. "I think somebody's in here."

La'ulu and I pushed our way through the broken glass and what remained of the door. We stood, bleeding and dripping wet, in front of two old women who were sitting on a small sofa, reading aloud from their Bibles.

"Are you O.K.?" I asked. "Do you have any injuries? Is there anyone else here?"

"It's just the two of us and we're fine," they replied. "We boarded up the house as best we could when the hurricane began."

"Do you have anything to eat?" I asked.

"Yes. Are you hungry?" they replied.

"No. I just wanted to make sure you're O.K. I think it's best that you stay here—you're above the waves, and if your house has withstood the storm this long, it will probably remain intact. The rest of the village is huddled in the school. We'll check on you as soon as the hurricane dies down."

The two women bid us farewell. La'ulu and I returned to the storm. It took us most of an hour to travel the first five hundred meters, but running and hiding behind large stones and tree trunks, we eventually made it back to the school. Barbara looked at my bleeding legs and chest.

"It must be terrible out there," she said.

"Yes. It's just too risky to search for any more survivors. We're going to have to wait until the storm dies down."

The next two days and nights passed like a long, unremitting nightmare. There was no relief from the storm. The sea did not retreat, though it seemed to stop rising. Each night, I kept my lonely vigil, facing the sea. We found an old radio one of the villagers had brought to the school. I rigged an antenna from some wire and replaced the batteries with those from my flashlight. After I repeatedly wiped sea spray from the battery contacts, the radio sputtered to life.

We listened intently to the AM band, but 2AP, the Western Samoa National Radio Service, was not on the air. One of the villagers said that they had been listening to the 2AP broadcast when they heard the announcer scream that a wave had hit the building, followed by silence. I wondered if we were the only people left alive in all of Samoa, but then I picked up, faintly, the Pago Pago station. The announcer asked everyone to remain indoors and said they had no word from Western Samoa. I turned to the shortwave band and managed to tune in the BBC World Service. The news reports highlighted the statements of European ministers on a unified currency and a political crisis in Pakistan, but there was no mention of a hurricane in the South Pacific. We were being destroyed, and no one would ever know.

It seemed crucial that we leave some sort of record, a testament of our struggle, in case we were all swept away to our deaths. I pulled my tape recorder out of the Pelican waterproof case and began interviewing the villagers huddled in the school, from the oldest to the youngest. What were their feelings about the storm? Most of the villagers seemed astonished by the storm's intensity, but were surprisingly optimistic about their chances of survival. On the recording, the whistling scream of the hurricane overshadows all of the conversations, which are barely audible. After finishing the tape, I placed it in the waterproof case with the antibiotics. If the wave did come, the case might be found still strapped to my body, and our families would know that we did not accept the inevitable without a fight, that we had struggled to the very last.

I found a pencil and began making some estimates of the hurricane's course in my notebook. The storm front seemed to be traveling at about six to seven miles per hour. Wind gusts must be in excess of two hundred miles per hour, the speed at which coconut trunks snap like pencils. The storm had hit us from the northwest, and the wind had not changed direction. Assuming that the storm system did not exceed five hundred miles in diameter, I calculated that we could expect a maximum of a hundred hours in the hurricane. We were approaching seventy-two hours—the third day. My estimates did not include the possibility that the hurricane might double back toward us, however, or that the storm might stop moving altogether and just sit on top of us for a week or more. If either of those things happened, our chances were slim. The school foundation was unlikely to survive another big wave, and if the tide began to rise again, we would all die.

The third night, after my turn at the wave watch, I fell into a deep sleep, holding Barbara very close. I awoke suddenly to an eerie silence and darkness. I thought we were dead. I couldn't hear anything. Slowly I struggled to my feet and stepped outside. The sky was black, but the wind had died. Incredibly, it appeared that the hurricane was over.

I went back to the mat and whispered in Barbara's ear.

"It's over. The wind is calm."

Barbara and I walked hand in hand to the door and looked toward the sea.

"Perhaps the eye of the hurricane is passing over us. Or perhaps the storm will come back," I said, not wanting to raise our hopes prematurely.

"No, Paul," Barbara said. "It's over."

Inside, I knew she was right. I noticed a dull yellow streak in the dark clouds, signaling the dawn's beginning. We went back inside the school and gazed at the sleeping forms of our four children. For the first time since the hurricane had started, Barbara began to weep. Hillary stirred and then opened her eyes. "Is that you, Daddy?" she asked.

"Yes, it's me, Hillary," I said, lifting her up into my arms. I carried her out of the room and held her in my lap in the broken chair in front of the school. Slowly the darkness began to recede, and then, from above the broken and tattered forest, the first ray of the sun shot from above the horizon. I sat, lacerated and in tattered clothing, holding my infant daughter. I looked through the open doorway at Emily, our teenager, at Paul Matthew, our son, and at Mary, the three of them sleeping peacefully. Nearly everything we possessed had been washed away, but I didn't care—my family was still alive. Everything we had worked for—the forest, my research project, the village, the school—all had been blown away. I had no idea how we would find enough to eat or how we would ever escape from Falealupo. But all that mattered to me at that moment was the sensation of Hillary's heartbeat. As I held her, tattered and bruised but still sleeping, and looked past her at the sunrise, I realized that never before had I seen anything so exquisitely beautiful. Never before had I been so grateful for the sheer joy of being alive. I knew that I now had deep responsibilities to the village, but I did not fear them. I had finally become human. Nafanua had returned.

B

A

10 cm.

Macropiper sp.

CHAPTER 15

Redemption

I leave Sisyphus at the base of
the mountain. One always finds
one's burden again. . . . The
struggle to the heights is enough
to fill a man's heart. One must
imagine Sisyphus happy.

Albert Camus,
The Myth of Sisyphus

The lulling hush of tropical rain on the tin roof draws my thoughts away from the small circle of light and conversation out toward the darkness. Pela has gone to bed, but Susi La'ulu sits quietly at the table with a cup of cocoa, having hung the mosquito nets. Lilo Manuele sits opposite me on a green chair, slapping mosquitos with a woven fan, watching me type on my laptop computer. "What happened to the book you were writing?" he asks.

"This is it," I reply.

I pause to chat with Lilo Manuele about the hurricane that destroyed the village and nearly killed all of us eight years ago. "The hurricane seems like a distant dream," Lilo Manuele says. "Sometimes people think about it, but the memories have faded."

Earlier that afternoon, after I again visited the stark remains of the Catholic church, a single wall and a steeple left like a grim relic of Hiroshima, I did not wonder why the villagers were reluctant to return to the former site of the village. I asked Tai'i then if he would someday rebuild his *fale* by the sea.

"No. None of the old people will. We are afraid of the wind and the wave." Tai'i was not exaggerating: Pela refuses even to inspect the site of her former home. *"Ua ou fefe i le galu"* ("I'm afraid of the wave"), she says quietly, gazing at the grave of her husband Lilo outside the new, tin-covered *fale.*

Given that Hurricane Ofa had struck precisely on the one-year anniversary of the signing of the rain forest covenant and the bestowal of the Nafanua title, I was surprised that the villagers failed to connect the events. Did the village incur God's wrath by conferring the title of an ancient deity? Was signing a covenant with foreigners a betrayal of village responsibility to preserve unfettered sovereignty over village lands? But none of the villagers ever raised these issues, perhaps because Hurricane Ofa was not a calamity limited to Falealupo. Though only eight people were killed by wave and wind throughout Samoa, damage totaled over 150 million U.S. dollars, a staggering sum for a country that is among the world's poorest. And although Falealupo was by far the hardest hit village (the newspaper *Savali* reported that "much of the village had either disappeared or been buried by sand"[59]), even distant Apia was battered by fifty-foot-high waves. And, perhaps most importantly, there was spiritual closure. Catholic Cardinal Pi'o Taofinu'u personally conducted a special service to rebury the human remains that had been uprooted from the graveyard by the waves.

I spent this morning listening to the children sing in the new school built inland by the forest. I helped the villagers to hammer an aluminum sign to it: "Falealupo Rain Forest School." After attaching the sign, I joined my research team in the rain forest behind the school, where Stephanie Hughes, pretty, petite, and three months pregnant, coached me through an arduous ascent up the fixed lines to a height of nearly thirty meters on a towering banyan tree. She then ascended the ropes herself in half the time it had taken me. The Seacology Foundation, a nonprofit organization I helped found, built the walkway with a generous donation from Nu Skin International, a company that markets personal care products based on plants used by indigenous people. Although the walkway will be an important resource for botanical research, our hope is that Falealupo will receive a small but constant stream of revenue from visitors to the walkway, each of whom will pay a modest admission fee. Village youths serving as rain forest guides could acquire skills in English or German, but with the chiefs firmly in control, cultural erosion should be minimal.

After catching my breath at the top of the huge banyan tree, I rappelled down to a branch where Stephanie's husband, Australian Kevin Jordan, sat tightening a stainless steel cable strung from the top of a *magaui* tree twenty-eight meters distant. Clipping a carabiner from my climbing harness to a piece of nylon webbing attached to the banyan tree, I examined a trailing vine, *Hoya samoensis*, that clung to the banyan high above the forest floor. Inspecting the flowers with a hand lens, I wondered if this vine was the *mafu'a* vine, the medicinal plant Pela had described to Dr. Kimberly Johnson and me the day before. During our interview, Kimberly had shown Pela a color photograph of herpes cold sores, and Pela had described the vine she uses to treat them. Since the National Cancer Institute declared prostratin, the molecule isolated from *Homalanthus*, a drug candidate for AIDS, I have been pursuing all possible Samoan antiviral leads.

I looked beyond a dangling *'ava'ava'aitu* vine at the Falealupo rain forest. The damage from Hurricane Ofa and the subsequent Hurricane Val is still apparent: several of the canopy trees stand leafless, having never recovered. I turned toward the other terminus of the nascent canopy walkway and gazed beyond a blackened swath stretching clear to Papa village, three miles distant, where a forest fire began during the dry season.

As I walked through the rain back toward Lilo Manuele's house, I thought of Falealupo in Ofa's wake. The image of the village after the hurricane was etched into my mind—no shade, repressive heat, and not a single structure remaining. A wasteland. On Sunday, the day the hurricane ended, we made a count of the villagers. It was then that we discovered that while most villagers had been wounded, some seriously, only one person had died, an elderly man, most likely from shock. Trunks of downed coconut trees littered the ground. Everything was topsy-turvy: concrete water tanks rested in the tops of trees, fish lay rotting in the forest, roofs of houses were pummeled by the waves on the edge of the reef, while large coral heads rested where the house foundations had once been. Most of the trees in the village had been blown down, and the few that remained were entirely leafless. The heat, made even more relentless by the absence of any shade, the smell of rotting animal carcasses, and the lack of drinking water and food, assaulted us after the storm receded.

It took us several days to clear the dirt track inland to the main road. The first two days we worked solely with hand tools, but on the third, Mark Muckerheide, the Peace Corps volunteer, arrived from Sataua with a chain saw he had borrowed from the logging company. Even then the work was difficult, but some of the chiefs were remarkably diligent in this task.

Late one afternoon, after the other villagers had ceased their labors, I found Fuiono Mase'ese'e' struggling to push a massive boulder from the road. I joined him, and with some effort, we finally succeeded in moving it to one side.

"Why are you working alone?" I asked him. "The other villagers have already left."

"They were tired. I have some strength remaining, so I thought I would continue."

Together, Fuiono Mase'ese'e and I worked until dark.

After I had done everything I could for the village in the storm's immediate aftermath, I told them that my family and I would try to leave Savaii. I wanted to move my family and John Cannon out of the disaster area and to raise funds for supplies desperately needed by the village. Although my van had been destroyed in the waves, Seumanutafa Nu'umau's truck, with the painted cow horns still affixed firmly to its hood, had survived inland. I prevailed upon Seumanutafa to lend it to us. With the fuel tank indicator on empty, we prayed the truck eighteen miles to Sala'ilua, where I offered twenty U.S. dollars to a puzzled fisherman for five gallons of outboard motor fuel. Asovao, the driver, was concerned that the oil-rich fuel might damage the truck engine, but he quieted when I told him the alternative was filling the tank with coconut oil. We finally made it to the Salelologa wharf, where we secured passage on the first boat to leave Savaii: the police boat *Nafanua*. The seas were very rough, but the captain gave us his cabin.

During our journey to Salelologa, I discovered that Falealupo was the worst hit of all the villages, although there was a great deal of destruction on the entire western part of Savaii. The airstrip at Asau had been washed out to sea, and the Vaisala Hotel had been severely damaged. Churches in Papa, Falelima, and Sataua were completely destroyed, and I estimated that over half of the population of western Savaii was left homeless.

The Mulifanua wharf had been so severely damaged that the *Nafanua* sailed past it and headed for Apia harbor. After docking, I was invited to a meeting of local leaders of the Mormon Church, where we composed a message requesting food and medical supplies from Church headquarters. Within two days, a cargo plane filled with the needed materials was dispatched to Samoa, landing slightly ahead of a similar aircraft sent by the Red Cross. In Apia I also met with Falealupo's Member of Parliament, urging him to send emergency food to the village via helicopter. Catholic Cardinal Pi'o, who had been born in Falealupo, was able to exert pressure on the government to dispatch heavy equipment to repair the village road.

After three days, I left Barbara and the children in the care of the Bethams and caught the first commercial flight out of Samoa to New Zealand. The Western Samoan ambassador used my photographs of the storm to buttress her plea for disaster assistance from

New Zealand. For a week I helped coordinate emergency relief efforts on behalf of the Western Samoan government during the day. Spending evenings on the telephone, I raised $75,000 from friends and family, with the bulk coming from Rex Maughan's company, the Swedish Nature Foundation, and Thomas Elmqvist and Bo Landin, to rebuild the Falealupo school and feed the village. Returning to Savaii from Apia two weeks later on a chartered aircraft with my son and Swedish filmmaker Bo Landin, I drove into Falealupo bearing two tons of rice, canned goods, chain saws, and tools.

"Nafanua, we thought you would never return," Tai'i said.

"I will always return," I replied.

And I have returned faithfully every year, to watch the forest heal and the village slowly rebuild. The forest has almost completely recovered, and the flying fox population is beginning to increase. Only a trained observer would realize that the forest was once completely denuded and left leafless by a once-in-a-century storm. Yet even now, eight years after the hurricane, none of the villagers have attempted to rebuild their *fale* where the waves swept the former village out to sea.

Next morning, as Fuiono Senio and I inspected work on the aerial walkway and then walked together through the forest, the hurricane seemed far in the past, but we did invite the American and Australian film crew that had come to the island on this trip to inspect the site of the Catholic church by the sea. Bo Landin's film documentary on the forest preserve and the aftermath of the hurricane had already been broadcast in Australia, Europe, and North America several years before. But the filming today was for a purpose that Fuiono and I were asked to keep confidential.

"How are you feeling?" I asked Fuiono when we returned with the film crew from the forest. I noticed that he had lost weight and occasionally stopped to catch his breath.

"My stomach hurts," Fuiono replied, an extraordinary admission for a Samoan who had been taught from infancy not to express pain.

Three weeks later, I sat next to Fuiono Senio in the front row of the Herbst Theater in San Francisco, a large Art Deco hall where the charter of the United Nations had been originally signed. Fuiono was dressed in a tapa print lavalava and equipped with an orator's staff and whisk. The lights dimmed, a large screen was lowered from the ceiling, and we watched the five-minute clip that had been filmed the previous month in Savaii. In abbreviated form, the film described our efforts to protect the Falealupo rain forest. Then

Fuiono and I walked up the stairs onto the stage to accept the $75,000 Goldman Environmental Prize from the heir to the Levi Strauss fortune, Richard Goldman.

Holding his staff in front of him, Fuiono ceremonially swished his orator's whisk three times over his shoulder. He then began speaking in chief's language to the nine-hundred-member audience: "E muamua ona ou fa'apa'i mālū atu i le pa'ia ma le mamalu o e ua fa'ae'e ma fa'apa'au i le maota." "I first gently address the sacredness and dignity of those assembled in this hall."

> At this time we praise and worship the love of God because tonight we greet each other in felicity, particularly on this occasion that we have anticipated being together.
>
> I first express my gratitude to Richard Goldman and the Goldman Foundation, because of his warrior spirit and his courage that allows him to raise the preservation of undisturbed forests as an issue for the entire world. I believe that the forest is the very work and artistry of God, created to give an abundant life to His people as well as the fowls of heaven.
>
> Therefore, I wish to express thanks and congratulations to Richard Goldman and the Goldman Foundation because of their efforts to preserve the planet. Our village has greatly added to forest preserves because we believe it is of importance for human life
>
> We also tender our gratitude for the high honor that you, sir, have awarded us this evening. Thank you.

The audience warmly applauded Fuiono's words. I then took the podium.

> Every day indigenous people are forced to choose between protecting their forests and building schools, water supplies, and medical clinics. Because of the generosity of the Goldman Foundation, at least one village will never have to make that choice again. My share of the Goldman Prize is being matched by two companies, Nu Skin International and Nature's Way, allowing the Seacology Foundation to establish a permanent endowment in excess of $100,000 for the Falealupo rain forest. I wish to thank my colleagues in Seacology, my father and mother who raised me in state and national parks and who taught me the meaning of conservation, my children, who pray nightly that God will protect the forest, and Barbara, who has held my hand through it all. As Camus said, "The struggle to the heights is enough to fill a man's heart." My heart is full tonight. Thank you very much.

Fuiono then presented his orator's staff and whisk to Mr. Goldman, empowering him to speak with courage on behalf of the world's rain forests.

As we walked from the podium amid the applause for the greatest award either of us would receive in this life, I looked at Fuiono's labored steps and knew that he was in pain. At the reception for us afterward, I asked an usher for a glass of water so that Fuiono could take a pill. A few days earlier, I had received very bad news about Fuiono's condition, and the night before had asked Mr. Goldman's son, a physician, to prescribe some pain medication.

We flew together to New York City for a press conference. As we sat in a small anteroom next to the hall filled with reporters, Fuiono asked, "Do you think the doctors will be able to help me?" I told him the results of his liver biopsy. "It doesn't look good, Fuiono. You've got liver cancer."

Fuiono looked sad, but unshaken. "Can they operate on me? I can pay the hospital bill with my Goldman Prize."

I promised Fuiono that I would arrange for him to be carefully evaluated by a medical team in Utah. As I spoke with him, I tried to be positive, but inside I felt my heart breaking, just as it had when I learned of my mother's test results so many years before.

After a visit with the U.N. Undersecretary General and another press conference in Washington, D.C., I brought Fuiono back to my home in Utah. Dr. Kurt Bodily, a gastroenterologist who, ironically, had once taken a biology class from me, asked me and Samoan Minister of Justice Solia Papu Va'ai (who had accompanied us on our trip) to help Fuiono remove his clothes. I was shocked to see how Fuiono's flesh hung from his upper arms. His skin was yellowish. His dark Samoan tattoo, extending from breast to knee, was distended by his swollen abdomen. Something about Fuiono's dark eyes, his dark tattoo, and his vulnerability as he looked at me reminded me of a Samoan flying fox.

Dr. Bodily carefully examined Fuiono with a stethoscope, pressing his liver and abdomen to discover regions of sensitivity. "Let's send him over for an immediate CAT scan," Dr. Bodily said.

After helping the technician position Fuiono on the sliding table, I retreated behind the lead-coated glass. I watched as successive computer-generated slices of Fuiono Senio appeared on the monitor.

"Look here," radiologist Dr. Gary Watt said, pointing to one. "The left lobe of the liver is enlarged, and both lobes are riddled with cancer. And here," he said, "we can see that the cancer has seeded the peritoneal cavity. No wonder his abdomen is swollen."

"Would an operation to remove a tumor from his stomach spare this man some pain, even if it was just for a few weeks?" I asked.

"There's no evidence of any physical obstruction to the gastric area," Dr. Watt replied. "In fact, we don't know where the tumor is. We probably could find it only during the autopsy."

"What about chemotherapy?"

"It won't help and will only make him sick."

"So there is nothing you can do?" I asked.

"The disease process is so advanced that I don't think chemotherapy or radiation would even function as a palliative. If I were your friend, I would just go home."

The drive with Fuiono to the Salt Lake City airport seemed to last forever. I tried to drive as smoothly as possible, to avoid any bumps or jars that could cause Fuiono further pain. Solia, Fuiono, and I stood together at the airline terminal. We gently embraced. Fuiono rested his head against mine. He seemed awkward without his lavalava, dressed in jeans and a sweater to keep out the chill.

"Goodbye, Fuiono," I said quietly in Samoan. "I love you." He left, without turning to wave. "I love you," I said, as he walked toward the plane that would carry him to where the sun sets in the sea, to the very gates of Pūlotu, back to Falealupo.

Acknowledgments

What happened to our family in Samoa, and the change it produced in my life, was overwhelming. I went to Samoa after my mother's death to try to find a cure for cancer and instead ended up watching a beloved Samoan friend die from the same disease— Fuiono Senio passed away in Falealupo a month after receiving the Goldman Prize. I was determined to maintain a low profile during my ethnobotanical studies in Samoa, but returned bearing the title of a native deity. I hoped to document the oral traditions and legends of a village whose name I picked off a map, and ended up watching the village I learned to love swept away by the waves.

Writing this book has not been easy—I am, after all, a botanist, rather than a novelist—but my primary goal has been to depict the deep humanity of the Samoan people. However, in too much of this narrative, I am at the center of events, when by rights, the story should not focus on a *palagi* botanist and his family, but instead on the remarkably courageous villagers of Falealupo. I hope that one day the villagers will produce their own written account.

I am very grateful to Chris and Claudia Cannon for the use of their Sundance cabin, where I began the emotionally wrenching task of writing a first draft. Subsequent drafts were produced in the Tetons, where I spent many enjoyable months of my boyhood.

I have read extended portions of the manuscript to Lilo Manuele and Lamositele Malofau, who made specific corrections on points of mythology. "We wouldn't want the world to think that we don't know our own legends," Lilo commented. I have also read portions to former Samoan Prime Minister Tuiatua Tupua Tamasese Efi and Member of Parliament Tupola George Hunt, who offered advice on Samoan customs. And the entire manuscript benefited from a careful and empathetic reading by Samoan writer Daniel Pouesi.

The scientific studies described in this book were supported by a Presidential Young Investigator Award from the National Science Foundation, the National Cancer Institute, the Institute for Polynesian Studies, the Schering-Plough Research Institute, and the University of Melbourne. In all of these studies, I am deeply indebted to my colleagues Drs. Thomas Elmqvist, Elizabeth Pierson, William Rainey, and Lars Bohlin, and many students. I am grateful to Brigham Young University for the academic freedom to pursue such an unusual line of research.

At W. H. Freeman, I am indebted to Mary Shuford for her initial positive reaction to the manuscript and for passing it on to Jonathan Cobb who added significant editorial assistance, encouragement, and patience. I also thank Marilyn Asay, Lorraine Clark, Brandon Dial, Francine Bennion, Thomas Elmqvist, John Enright, Christopher Hallowell, Neal Kramer, Bo Landin, Rebecca Marshall, Cory Maxwell, Clark and Patty Monson, Deirdre Paulsen, Dixie Pierson, Bill Rainey, Norma Roche, the Utah Arts Council, Amber Scott, and Sandie Tillotson for reading portions of the manuscript.

Verne and Marion Read, Ken Murdock, Rex Maughan, The Goldman Environmental Foundation, Nature's Way Nu Skin International, Betty Sung, Sandie Tillotson, and many donors to the Seacology Foundation have helped Falealupo village protect its forest. You can obtain more information on the village-based conservation preserves in Samoa, as well as copies of Bo Landin's documentary film based on the events described in this book, by writing the Seacology Foundation at Box 340, Lawai, Kauai, Hawaii 96765 or on the Internet at http://www.seacology.org.

I thank Emily, Paul Matthew, Mary, and Hillary for their courage in Samoa and for their support of conservation. Perhaps someday their story can be told for the benefit of their new sister Jane. And most of all I thank Barbara, who is the love of my life.

Spring Creek Ranch, Jackson Hole, Wyoming

Sources for Epigraphs

CHAPTER 1
Albert Camus, *Notebooks 1935–1942* (New York: Alfred A. Knopf, 1963), 13–14.

CHAPTER 2
W. Somerset Maugham, *A Writer's Notebook* (New York: Penguin Books, 1967), 118.

CHAPTER 3
Robert Louis Stevenson, "The Beach at Falesa," in *The Short Stories of Robert Louis Stevenson* (New York: Charles Scribner's Sons, 1935), 435.

CHAPTER 4
Joseph Conrad, "The Return," in *Tales of Unrest* (Garden City: Doubleday, Page & Co., 1922), 205.

CHAPTER 5
Paul Gauguin, *Noa Noa: The Tahiti Journal of Paul Gauguin*, trans. O. F. Theis (San Francisco: Chronicle Books, 1994), 37.

CHAPTER 6
Albert Camus, *The Myth of Sisyphus and Other Essays*, trans. Justin O'Brien (New York: Random House, 1991), 123.

CHAPTER 7
Jean-Paul Sartre, *Nausea*, trans. Lloyd Alexander (New York: New Directions, 1964), 96–98.

CHAPTER 8
Herman Hesse, *Steppenwolf*, trans. Basil Creighton (New York: Bantam, 1969), 24.

CHAPTER 9
Joseph Conrad, *Lord Jim* (New York: Penguin, 1980), 182.

CHAPTER 10
W. B. Yeats, "Responsibilities," in *W. B. Yeats: The Poems*, ed. Daniel Albright (London: J. M. Dent & Sons, 1990), 148.

CHAPTER 11
Sartre, *Nausea*, 5.

CHAPTER 12
Albert Camus, "The Plague," in *The Collected Fiction of Albert Camus* (London: Hamish Hamilton, 1960), 250.

CHAPTER 13
Rudyard Kipling, "The Sea and the Hills," in *The Collected Verse of Rudyard Kipling* (New York: Doubleday, Page & Co, 1910), 23–25.

CHAPTER 14
Marcel Proust, *Le Temps Retrouvé* (Time Regained), trans. S. Hudson (London: Chatto & Windus, 1956), 215.

CHAPTER 15
Camus, *The Myth of Sisyphus and Other Essays*, 20.

Notes on
Illustrations

All illustrations, other than the steel engravings drawn by A. T. Agate and engraved by Alfred Jones for the 1849 *Narrative of the U.S. Exploring Expedition,* were drawn by Michael Rothman. The Rothman drawing in Chapter 5 is of the Samoan flying fox, *Pteropus samoensis.* In the following list of the drawings at the beginning of each chapter, each plant is identified by Latin binomial, author, and herbarium voucher reference number.

FRONTISPIECE
 Homalanthus nutans Guill. (Cox 842)
 Samoan name: *Mamala*
Water infusions of the inner bark are used to treat hepatitis and *tulitā,* an internal ailment of the abdomen. After Epenesa's death, her daughter Malama told me that there are two types of the tree, distinguished by petiole color; only one of these is used for hepatitis. The anti-AIDS drug candidate prostratin was isolated from *Homalanthus nutans.*

CHAPTER 1
 Hibiscus tiliaceus Linn. (Cox 984)
 Samoan name: *Fau*
 English name: Hibiscus
 A. Habit
 B. Unopened flowers
 C. Androgynophore
 D. Placement of anthers on androgynophore
 E. Anthers with pollen
Healers use the bark to treat appendicitis. The sap of the roots is dripped into the eyes to treat eye injuries. The plant is also used as an ornamental in village gardens.

CHAPTER 2
Piper methysticum Forster f. (Cox 1065)
Samoan name: *Kava*
Kava is the ceremonial drink of Polynesia. Its slight sedative effect is produced by lactones and sesquiterpenes, tranquilizing molecules in the plant's underground stem. Given its pharmacological activity, it is odd that kava is rarely used in traditional medicine. Kava is sometimes drunk after a long day of working in the sun to relieve sore muscles.

CHAPTER 3
Syzygium malaccense (L.) Merrill & Perry (Cox 841)
Samoan name: *Nonu fi'afi'a*
Healers use the leaf galls to treat an internal disease called *mūmū tuaula*, well known to the early missionaries.

CHAPTER 4
Artocarpus altilis (Parkinson) Fosberg (Cox 1036)
Samoan name: *Ulu ea*
English name: Breadfruit
The roots are used to treat oral thrush and fungal infections of the anus. The volleyball-sized fruits are roasted and eaten as a staple throughout the islands.

CHAPTER 5
Atuna racemosa Rafin. (Cox 1035)
Samoan name: *Ifiifi*
The fruits of this tree are an important component of Samoan oil. Several anti-inflammatory compounds have been isolated from the fruits sent by me to researchers at Uppsala University.

CHAPTER 6
Scaevola taccada (Gaertn.) Roxb. (Cox 1022)
Samoan name: *To'ito'i*
The bark of this seaside shrub is used to treat irregularities of menstruation, and the roots are used in cases of fish poisoning.

CHAPTER 7
Inocarpus fagifer (Parkinson) Fosberg (Cox 1030)
Samoan name: *Ifi*
Water infusions of the bark are used to treat appendicitis and stomach pain, and the roots are used to treat the gastric illness *fe'efe'e*. In Tahiti the tree often grows near ruined temples.

CHAPTER 8
   Cananga odorata Hook. F. & Thoms. (Cox 975)
   Samoan name: *Moso'oi*
Water infusions of the bark are drunk as a treatment for asthma. The flower petals are used to scent Samoan oil and to manufacture fragrant leis for ceremonial occasions.

CHAPTER 9
   Piper graefii Warburg (Cox 824)
   Samoan name: *Fue manogi*
The bark is used to treat appendicitis and stomach pain. The stems are used to treat the disease of the *toala* (the center of being beneath the navel) called *oso fa puni moa*—an abdominal blockage caused by *toala* movement, producing symptoms similar to those caused by kidney stones. The leaves are used to treat the inflammatory disease *mūmū afi*. This woody liana grows only in the primary rain forest.

CHAPTER 10
   Ipomoea pes-caprae Roth. (Cox 853)
   Samoan name: *Fue moa*
   English name: Beach morning glory
The leaves of this beach plant are used to treat inflammation and the disease *mūmū lele*.

CHAPTER 11
   Psychotria insularum A. Gray (Cox 1055)
   Samoan name: *Matalafi*
   A. Habit
   B. Fruit with persistent calyx
   C. Schematic of fruit with single ovule
   D. Seed
   E. Cross-section of seed
This tree is considered by traditional healers to be one of the most potent in Samoa. Leaves are used to treat complications of childbirth, boils, intestinal distress, and many types of *mūmū;* the young leaves are used to treat acne and cataracts; and the stems are used to treat the *toala* disease called *oso fa puni moa.*

CHAPTER 12
   Callophyllum inophyllum (Cox 1004)
   Samoan name: *Fetau*
   A. Habit
   B. Transection of fruit

The leaves are used to treat the inflammatory disease *mūmū afi*. Kava ceremonies recount the legend of the Samoan princess, Leutogi Tupa'itea, who was saved in a *fetau* tree. Angry with her infertility, the King of Tonga drove her into the forest, where she hid in the crotch of a *fetau* tree. Angry villagers piled wood at the base and lit the fire. But just as the flames began to rise, the princess was miraculously saved by a flight of flying foxes, leading to the saying *"Ua tatou maga fetau soifua,"* "We have had life in the crotch of the *fetau* tree." The fetau tree is thus a symbol of hope in Samoa.

CHAPTER 13
  Colubrina asiatica Brongn. (Cox 985)
  Samoan name: *Fisoa*
  A. Habit
  B. Axillary inflorescence
  C. Lateral view of fruit
  D. Dorsal view of fruit
  E. Flower
  F. Lateral view of fruit
  G. Flower with hypogenous disk and stamens
The leaves are used for complications of childbirth. Anciently the saponin-rich plant was used for washing with large quantitites of water.

CHAPTER 14
  Erythrina variegata Linn. (Cox 1062)
  Samoan name: *Gatae*
Healers use the bark topically to treat inflammation, but only the Samoan variety *gatae palagi* is said to be efficacious. Paul Matthew was saved from a catastrophic encounter with a wasp nest by this bark.

CHAPTER 15
  Macropiper sp. (Cox 830)
  Samoan name: *'Ava'ava'aitu*
  A. Stem habit with swollen nodes
  B. Closeup of infructescence
The stem and leaves of the plant are used to treat the inflammatory disease *mūmū lele*. The Samoan name means "kava of the ghosts," but the reasons for this have been lost.

Notes

1. John B. Stair, *Old Samoa* (London: The Religious Tract Society, 1897), 218–19.

2. Charles Wilkes, *Narrative of the United States Exploring Expedition during the Years 1838, 1839, 1840, 1841, 1842,* vol. II (Philadelphia: Lea and Blanchard, 1845), 131.

3. George Turner, *Samoa, A Hundred Years Ago and Long Ago* (London: Macmillan, 1884), 39.

4. Ibid., 40.

5. Stair, *Old Samoa,* 248.

6. Turner, *Samoa, A Hundred Years Ago,* 39–40.

7. John Williams, *A Narrative of Missionary Enterprises in the South Seas* (London: John Snow, 1838), 283.

8. Ibid., 284.

9. Ibid., 286–87.

10. Ibid., 287–88.

11. Edward Edwards and George Hamilton, *Voyage of H.M.S. "Pandora"* (London: Francis Edwards, 1915), 1–49.

12. David Steadman, "Biogeography of Tongan Birds Before and After Human Impact," *Proceedings of the National Academy of Sciences* 90 (1993): 818–22.

13. T. H. Hood, *Notes of a Cruise in H.M.S. "Fawn" in the Western Pacific in the Year 1862* (Edinburgh: Edmonston and Douglas, 1863), 34.

14. H. J. Moors, *With Stevenson in Samoa* (Boston: Small, Maynard & Company, 1910), 10.

15. Joel M. Hanna, David L. Pelletier, and Vanessa J. Brown, "The Diet and Nutrition of Contemporary Samoans," in *The Changing Samoans: Behavior and Health in Transition,* ed. Paul T. Baker, Joel M. Hanna, and Thelma S. Baker (Oxford: Oxford University Press, 1986), 281.

16. D. Arnold, "Introduction: Disease, Medicine, and Empire," in *Imperial Medicine and Indigenous Societies,* ed. D. Arnold (Manchester: Manchester University Press, 1988), 3.

17. Megan Vaughan, *Curing Their Ills: Colonial Power and African Illness* (Stanford: Stanford University Press, 1991), 161–62.

18. Ibid., 165.

19. Turner, *Samoa, A Hundred Years Ago,* 139.

20. T. Trood, *Island Reminisces* (Sydney: McCarron, Stewart, and Co., 1912), 65.

21. A. B. Steinberger, "Message from the President of the United States," 44th Congress, House of Representatives, Executive Document 161 (1876), 1–125.

22. D. Hunt, "Samoan Medicines and Practices," in *U.S. Naval Medical Bulletin* 19 (1923): 145–52.

23. J. F. G. de La Pérouse, *A Voyage Round the World, Performed in the Years 1785, 1786, 1787, and 1788 by the Boussole and Astrolabe,* vol. III (London: J. Johnson, St. Paul's Churchyard, 1798), 104–5.

24. Edwards and Hamilton, *Voyage of H.M.S. "Pandora,"* 130.

25. Wilkes, *Narrative of the United States Exploring Expedition,* 125.

26. John Erskine, *Journal of a Cruise among the Islands of the Western Pacific* (London: John Murray, 1853), 51.

27. T. H. Hood, *Notes of a Cruise in H.M.S. "Fawn,"* 44.

28. La Pérouse, *A Voyage Round the World,* 63.

29. Jacob Roggeveen, *The Journal of Jacob Roggeveen,* ed. Andrew Sharp (Oxford: Clarendon Press, 1970), 151.

30. La Pérouse, *A Voyage Round the World,* 90.

31. Williams, *A Narrative of Missionary Enterprises,* 291.

32. Wilkes, *Narrative of the United States Exploring Expedition,* 126.

33. Erskine, *Journal of a Cruise,* 86.

34. Stephen Allen, quoted in Michael J. Field, *Mau: Samoa's Struggle for Freedom* (Auckland: Polynesian Press, 1984), 200–201.

35. Margaret Mead, *Coming of Age in Samoa* (New York: William Morrow, 1928), 14–15.

36. Ibid., 16.

37. Ibid., 130–31.

38. Ibid., 162.

39. Wilkes, *Narrative of the United States Exploring Expedition,* 73.

40. Ibid., 125.

41. Ibid., 138.

42. Derek Freeman, *Margaret Mead and Samoa: The Making and Unmaking of an Anthropological Myth* (Cambridge: Harvard University Press, 1983), 227–53.

43. Ibid.

44. Derek Freeman, quoted in *Honolulu Star-Bulletin,* 9 March 1983, B-1.

45. Ibid.

46. Joseph Conrad, *Lord Jim* (New York: Penguin, 1980), 182.

47. Peale R. Titian, "Zoology," in *U.S. Exploring Expedition,* vol. 8 (Philadelphia: C. Sherman, 1848), 5.

48. Ibid., 21.

49. K. Anderson, *Catalogue of the Chiroptera of the British Museum* (1912), 287.

50. Paul Shankman, "Notes on a Corporate 'Potlatch:' The Lumber Industry in Samoa," in *The World as a Company Town: Multinational Corporations and Social Change,* ed. Ahamed Idris-Soven, Elisabeth Idris-Soven, and Mary K. Vaughn, papers prepared for the 19th International Congress of Anthropological and Ethnological Sciences, Chicago 1973 (The Hague [Noordeinde 41]: Mouton, 1978), 387.

51. Annie Dillard, *Pilgrim at Tinker Creek* (New York: Harper's Magazine Press, 1974), 112.

52. Paul Gauguin, *Noa Noa: The Tahiti Journal of Paul Gauguin,* trans. O. F. Theis (San Francisco: Chronicle Books, 1994), 37.

53. This and all other translations of Samoan texts are my own.

54. W. B. Churchward, *My Consulate in Samoa* (London: Bentley, 1887), 140.

55. Theodore Roethke, "The Waking," in *The Collected Verse of Theodore Roethke: Words for the Wind* (Bloomington: Indiana University Press, 1961), 124. Used with permission.

56. James Cook, *The Journal of the H.M.S. Endeavour 1768–1771* (Surrey, England: Genesis Pub. Ltd. and Rigby Ltd., 1997), 19.

57. Erskine, *Journal of a Cruise,* 111.

58. "Fleeing the Water's Edge." *Coastal Heritage* 12 (1997): 10–12.

59. *Savali,* March 26, 1990.

About the Author

Named by *Time* magazine as a "Hero of Medicine," ethnobotanist Paul Alan Cox is the Director of the U.S. National Tropical Botanical Garden, a system of Congressionally chartered gardens throughout Hawaii and Florida. He also serves as King Carl XVI Gustaf Professor of Environmental Science at the Swedish Biodiversity Centre in Uppsala, Sweden. For his conservation efforts in Samoa, Cox was awarded the Goldman Environmental Prize in 1997. Considered the "Nobel Prize of the Environment," the Goldman Prize is awarded annually to an environmentalist from each of the six inhabited continents.